To Ric & P.J.,
Good friends,
best of neighbors.
Long live the beach!
Kristin

Christianity in a Time
of Climate Change

Christianity in a Time of Climate Change

To Give a Future with Hope

Kristen Poole

WIPF & STOCK · Eugene, Oregon

CHRISTIANITY IN A TIME OF CLIMATE CHANGE
To Give a Future with Hope

Image permissions:
A cedar tree in the Maasser Cedar Forest, south of Beirut. JOSH HANER/The New
York Times/Redux. Reproduced with permission of The New York Times/Redux.
Raymond L. Lindeman, "Seasonal Food-Cycle Dynamics in a Senescent Lake,"
American Midland Naturalist 26.3 (1941) 637. Reproduced with permission of
American Midland Naturalist.
Cloud of Monarch Butterflies. ©Ingo Arndt/naturepl.com. Reproduced with permis-
sion of the Nature Picture Library.
Brueghel, Pieter the Elder (c. 1525–1569). *Landscape with the Fall of Icarus*. Scala/
Art Resource, NY. Reproduced with permission of Art Resource.

Wipf & Stock
An Imprint of Wipf and Stock Publishers
199 W. 8th Ave., Suite 3
Eugene, OR 97401

www.wipfandstock.com

PAPERBACK ISBN: 978-1-7252-5713-9
HARDCOVER ISBN: 978-1-7252-5714-6
EBOOK ISBN: 978-1-7252-5715-3

Manufactured in the U.S.A.   04/16/20

For Corinna and Juliana

If you remove the yoke from among you,
      the pointing of the finger, the speaking of evil,
      if you offer your food to the hungry
      and satisfy the needs of the afflicted,
then your light shall rise in the darkness
      and your gloom be like the noonday.
      The LORD will guide you continually,
      and satisfy your needs in parched places,
      and make your bones strong;
and you shall be like a watered garden,
      like a spring of water,
      whose waters never fail.
      Your ancient ruins shall be rebuilt;
      you shall raise up the foundations of many generations;
you shall be called the repairer of the breach,
      the restorer of streets to live in.

                    —Isaiah 58:9–12

Contents

List of Illustrations

Figure 1. A home flooded by Tropical Storm Erin in Kingfisher, Oklahoma, 2007; FEMA Photo Library/Marvin Nauman.

Figure 2. A cedar tree in the Maasser Cedar Forest, south of Beirut. JOSH HANER/The New York Times/Redux

Figure 3. *Renatus Des-Cartes excellent compendium of musick* (London, 1653). Folger Shakespeare Library.

Figure 4. Raymond L. Lindeman, "Seasonal Food-Cycle Dynamics in a Senescent Lake," *American Midland Naturalist* 26.3 (1941) 637.

Figure 5. Triquetra. Wikimedia Commons.

Figure 6. Jacob Peter Gowy, *The Flight of Icarus* (1635–1637). Museo del Prado, Madrid.

Figure 7. Cloud of Monarch Butterflies. ©Ingo Arndt/naturepl.com.

Figure 8. Pieter Brueghel the Elder, *Landscape with the Fall of Icarus* (ca. 1558). Royal Museums of Fine Arts of Belgium. Photo Credit: Scala/Art Resource, NY.

Figure 9. Hans Süss von Kulmbach, *The Ascension of Christ* (1513). The Metropolitan Museum of Art, Rogers Fund, 1921.

Acknowledgments

I would like to thank all of those people who guided, supported, and nurtured me through the process of writing this book.

This book marks the end of a journey. Many years ago, in a phase of my life that was already overflowing with career and family obligations, I got it into my head that I needed to take up the serious study of theology. As a scholar of the English Reformation, this impulse was intellectual, a desire to learn about the early theologians who shaped the thought of sixteenth- and seventeenth-century England. But there was another pull I couldn't quite name. Over the ten years it took me to complete a Master's of Sacred Theology (all the while raising children and continuing in my vocation as an English professor), many people asked me what I planned to do with the degree. I, too, wondered what I was supposed to be doing. At the very end of the degree program, after having spent most of my time studying the early years of the Christian church, it became surprisingly clear to me that writing this book—a project that looks not to the past, but to the future— was what I was supposed to do. Even though this book is on a topic I never covered in seminary, it is the culmination of a long period of study and reflection that was guided by many inspiring teachers. At United Lutheran Seminary (formerly the Lutheran Theological Seminary), Philadelphia, I benefitted from the knowledge, wisdom, and intellectual rigor of Drs. Erik Heen, Walter Wagner, and J. Jayakiran Sebastian (all of whom, in different ways, balance demanding seminars with a delightful sense of humor). Most especially, Dr. John Hoffmeyer has been a model of a deeply thoughtful and socially engaged theologian; through his classes and an independent study, I gained not just knowledge, but a spiritual approach to theological inquiry. I landed on the idea to write my Master's thesis on climate change in a seminar taught by Dr. Barry Bates, and I was grateful that Dr. Hoffmeyer agreed to serve as my thesis director. Dr. Crystal Hall was my second reader, and her generous comments on the thesis prompted my thoughts into new directions. The final book is a much-expanded version of the thesis, but it would not have taken shape without the initial guidance of Drs. Hoffmeyer and Hall.

At the University of Delaware, I could not have asked for a better department Chair than Dr. John Ernest. Teaching at an institution that is so secular as to not even have a religion department can be a little lonely for someone who specializes in religious history and literature, and John not only respected but encouraged my theological studies. This project marks a significant departure from my usual scholarship, and it was extremely helpful to feel supported by John's enthusiasm.

Many of the ideas for this book grew out of conversations with my friends at St. Paul's Episcopal Church in Chestnut Hill, Philadelphia, especially those who joined me to start up the Climate Change Reading Group. My fellow volunteers at the Citizens' Climate Lobby, those in the local Philadelphia chapter and in the national organization, have also been an inspiration for their dedication, perseverance, and optimism. I have found CCL meetings to be not only a positive way of taking climate action, but also intellectual and thought-provoking. If this book marks the end of one journey, it perhaps also marks the beginning of a new one. I used to think that addressing climate change was the responsibility of scientists, economists, and politicians, but now I realize that all of us have a role to play. As I discuss in the book's epilogue, I have tried to follow Paul's exhortation to discern one's gifts and use them for the common good. Between my church and CCL, I have been figuring out how I can use my talents in the mission to mitigate climate change and secure a livable world for our heirs.

My special thanks go to Dr. Matthias Ohr, as much a humanist as a scientist, who graciously read the book manuscript and helped me to iron out many wrinkles. My former Ph.D. student, Hannah Eagleson, served as my copyeditor and proofreader for the final manuscript. She read with a meticulous eye, and it was a pleasure to have the tables turned as I received her careful corrections and comments. In addition to catching my typographical errors, she also provided insightful feedback on the book from the point of view of an evangelical reader, for which I am very grateful.

At Wipf and Stock, Matthew Wimer has been a steady, cheerful editorial guide through the publication process. George Callihan efficiently steered the manuscript into production, while Savanah Landerholm expertly typeset the book. Naomi Linzer, who has now worked with me on four different volumes, once again produced a careful and thoughtful index.

My husband Martin Brückner has cheered me on and also carried on when I was busy writing in a room of my own; his help extended from reading chapter drafts to covering more than his fair share of household chores. Thank you.

This book is dedicated to, and motivated by, my beautiful daughters Corinna and Juliana, inheritors of the future earth.

List of Abbreviations

Col	Colossians
Cor	Corinthians
Dan	Daniel
Eph	Ephesians
Exod	Exodus
Gal	Galatians
Gen	Genesis
Jas	James
Jer	Jeremiah
Kgs	Kings
Lam	Lamentations
Matt	Matthew
Mic	Micah
OED	*Oxford English Dictionary*
Phil	Philippians
Ps	Psalms
Rom	Romans

Introduction

A New Ethics of the Neighbor

Is there anyone among you who, if your child asks for bread, will give a stone? Or if the child asks for a fish, will give a snake? ... In everything do to others as you would have them do to you ...

—MATTHEW 7:9-10, 12

OLD WISDOM, NEW PHYSICS

In the Bible, we often read of the end of the world, which is portrayed through vivid descriptions. Consider this passage, from Second Peter:

> But the day of the Lord will come like a thief, and then the heavens will pass away with a loud noise, and the elements will be dissolved with fire, and the earth and everything that is done on it will be disclosed. Since all these things are to be dissolved in this way, what sort of persons ought you to be in leading lives of holiness and godliness, waiting for and hastening the coming of the day of God, because of which the heavens will be set ablaze and dissolved, and the elements will melt with fire? But, in accordance with his promise, we wait for new heavens and a new earth, where righteousness is at home. (2 Pet 3: 10–13)[1]

1. Unless specified otherwise, all biblical citations in this book are from the New Revised Standard Version (NRSV) of the Bible, copyrighted 1989 by the Division of Christian Education and the National Council of the Churches in Christ in the United States of America, used with permission in Accordance Bible software, version 12.3.2.

This epistle expresses a particular scientific understanding of the material world, that of Stoic physics.[2] In the ancient world, there were two competing theories of matter. The Epicureans contended that matter was comprised of atoms that were in an endless process of re-arranging themselves to form the objects of the world. The Stoics disagreed, arguing that all matter was part of a fiery, fluid continuum known as *pneuma*, which was also the all-pervasive spirit of the universe. In the Stoic understanding, the universe experiences life cycles which culminate in a final dissolution of the cosmos through a conflagration known as *ekpyrosis*.[3] Christianity absorbed much from the precepts of Stoic physics: the notion of the *pneuma* became a way of speaking of the Holy Spirit,[4] and the *ekpyrosis* became a way of understanding the end of the world. This epistle of Second Peter, with its assertion that "the elements will be dissolved with fire" and that "the heavens will be set ablaze and dissolved, and the elements will melt with fire," thus articulates a common scientific theory of the period. That scientific theory, in turn, shaped the accompanying religious belief. Throughout time, the way in which we understand the physical properties of matter has informed our beliefs and the moral and ethical code that flows from those beliefs.

Recognizing the historical scientific perspective of Second Peter forces us to contemplate our understanding of this biblical text. While our current comprehension of physics shares elements of both Epicurean and Stoic philosophies (we understand the world to be made of discrete atoms, but in the realm of quantum mechanics these atoms can behave more like a fluid continuum than individualized particles), we no longer believe that the universe has life cycles that periodically end, phoenix-like, in a great conflagration that gives birth to a new cosmos. From a scientific perspective, many modern people would consider the physics of Second Peter to be wrong. How does that alter our reading of this passage? There is still deep spiritual wisdom to be found here. Our physical world is fragile; our own worldly existence is transient. In light of this fragility and transience, the central question of the passage is not only still relevant, but urgent: "what sort of persons ought you to be in leading lives of holiness and godliness?"

As it happens, 2 Peter 3:10 presents "one of the most difficult problems in the New Testament"[5] because of the Greek text that has been translated above

2. Harrill, "Stoic Physics," 115–40.

3. Sambursky, *Physics of the Stoics*, 106.

4. Lapidge, "Stoic Cosmology," 176.

5. W. Hall Harris, note on 2 Peter 3:10 in the New English Translation, copyright

as "the earth and everything that is done on it will be disclosed." What is meant by "disclosed"? Other translations render the text as "the earth and every deed done on it will be *laid bare*" (New English Translation), or "the earth also and the works that are therein shall be *burned up*" (King James Version).[6] These translations, more dire than the NRSV's "disclosed," can take us to a modern phenomenon that, if not quite an apocalyptic cosmic firestorm, shows us a world "laid bare" and "burned up": the recent rise of cataclysmic wildfires. We can look to the so-called "Camp Fire" or "Paradise Fire" of 2018, in which about 153,000 acres and almost 19,000 structures were destroyed in California, most of the destruction happening in the first four hours of the disaster.[7] The idea of a "camp fire" is a primal expression of hearth, of human community and collective safety; Paradise happens to be the name of the town destroyed by wind and fire, but "paradise" connotes the Garden of Eden, a God-given place of human and natural perfection. The two names of the fire thus ironically add to a sense of shattered security. Seeing the images of the damage, it does seem as if the earth and humanity were destroyed on a biblical scale.

The counsel of Second Peter in the face of an expected apocalypse, imagined as the Stoic *ekpyrosis*, is to "wait for new heavens and a new earth." There is wisdom in calling for patience as we experience the long arc towards the time when it shall be on earth as it is in heaven, a time when human hearts and actions align fully with the divine commandment to love God and love neighbor. But in scientific terms, today few of us expect a spontaneous fiery end and subsequent re-composition of the entire universe. Instead, we are witnessing—and participating in—a different kind of environmental destruction, that of global climate change. We now have a different knowledge of physics from those in the first centuries after Christ. We understand that the burning of fossil fuels like coal and oil—that is, the burning of ancient carbon, the residue of ancient life forms that decomposed deep within the earth—releases carbon dioxide (CO_2)[8] into the

---

1996–2005, by Biblical Studies Press, in Accordance Bible Software, version 12.3.2, note 34.

6. New English Translation, copyright 1996–2005, by Biblical Studies Press, in Accordance Bible Software, version 12.3.2; King James Version with Strong's Numbers (KJVS), public domain, formatted and corrected by OakTree Software, Inc., Version 3.5, in Accordance Bible software, version 12.3.2.

7. Wikipedia, "Camp Fire."

8. A quick gloss on the meaning of "CO_2" might be helpful for some readers. CO_2 is a chemical formula (also sometimes written CO2 for typographic convenience) that indicates a molecule formed from one atom of carbon attached to two atoms of oxygen. These molecules form an odorless, colorless gas.

atmosphere, and that by pumping more CO_2 into the planet's system than it can process, we are contributing to an atmospheric layer that is increasingly trapping the sun's heat, thereby warming the planet. This trapped heat is now rapidly altering weather patterns that have remained relatively stable for millennia. These new weather patterns are wreaking havoc with human and animal habitat, leading to catastrophic heatwaves and fires, floods and droughts, ice melt and (perhaps counterintuitively[9]) frigid winters as the environmental conditions around the world shift in unpredictable ways. And we don't yet know what we still don't know: given the massive scope of the problem, given that careful scientific studies often have a narrow focus and long duration, and given a rate of climate change that has far outpaced all but the most extreme scientific predictions,[10] that which we understand about how the environment will transform in the future is just the tiniest tip of the iceberg, to use a proverbial expression that is itself endangered by the accelerating rate of melting sea ice.

THE FUTURE OF THE GOLDEN RULE

Our planetary crisis is a different situation than waiting for God's Kingdom, and it requires us to do something other than simply waiting. There is an urgent need for action and an impetus for change: we need changes in technology, changes in policy, changes in collective behaviors. But perhaps above all we need to change our ethical outlook on the world. The Christian ethics that have developed alongside the growth of Western consumerism for the past few centuries are often at odds with the ethical considerations that are needed in the time of climate change. The environmental crisis presents humanity with a new, unique moral problem in terms of both space and time. In terms of space, the planetary effects of climate change defy the local logic of previous environmental concerns. If acid rain was killing the Black Forest in Germany, Germans (and their willing neighbors) could change the type of gasoline in their cars and reduce the other industrial pollutants that were the source of the problem. If the Chicago River was polluted, the people of Chicago and Illinois (and their willing neighbors) could stop dumping pollution in the river. But the sources of climate change cannot be localized:

9. For a quick and accessible account of how extremely cold winter temperatures are related to climate change, see Koren and Meyer, "Colder Than Mars"; Pierre-Louis, "Why Is the Cold Weather So Extreme" and "Brace for the Polar Vortex."

10. Linden, "How Scientists Got Climate Change So Wrong."

the emissions of a car in India, or Brazil, or Wyoming do not simply hang in the air over that space. Carbon dioxide, methane, and other greenhouse gasses do not know geopolitical boundaries. And these gasses do not know temporal boundaries, either. While greenhouse gasses can have varying life spans,[11] the fundamental problem—the human-caused transfer of carbon from earth to air—has been created over centuries, although this carbon transfer has escalated at breakneck speeds in the last decades with the rise of an increasingly mobile and global world.

What resources does Christianity offer for rethinking what it means to lead lives of holiness and godliness in a time of climate change? First, of utmost importance is the fact that Christianity is about personal and communal change. Not only can we change our own hearts, minds, and behaviors, but we are called to do so as a condition of genuine spiritual growth. Nicodemus's scoffing question to Jesus, "How can anyone be born after having grown old?" (John 3:4), assumes that people are set in their ways. Jesus's response, "no one can enter the kingdom of God without being born of water and Spirit" (John 3:5), establishes change as a condition of godliness, a change that is brought about by a combination of individual volition (the choice to be baptized, or to renew or live out one's baptismal vows) and God's spiritual help. Second, Christianity distills God's commandments to the essentials. In the Gospel of Mark we read of two commandments: "love the Lord your God with all your heart, and with all your soul, and with all your mind, and with all your strength," and "love your neighbor as yourself" (Mark 12:30, 31). And in John 13:34 Jesus says, "I give you a new commandment, that you love one another. Just as I have loved you, you also should love one another." Love God, love neighbor, love each other. Pretty simple, really.

Or not. In the Gospel of Luke, the synthesis of the law into "love the Lord your God with all your heart . . . love your neighbor as yourself" is articulated not by Jesus but by a lawyer who is testing him. As soon as Jesus affirms that this was the right answer to a question about inheriting eternal life, the lawyer promptly asks, "And who is my neighbor?" (Luke 10:29). We have a human tendency to look for loopholes in God's law, to quibble about the terms and to qualify our responsibilities. Jesus's answer to the lawyer's question does not provide a definition or a taxonomy of the neighbor, but

---

11. See Clark, "How Long Do Greenhouse Gases Stay." For CO2, the most significant human-made greenhouse gas, 65 percent–80 percent can be dissolved in the ocean over twenty to two hundred years, while "the rest is removed by slower processes that take up to several hundreds of thousands of years."

offers instead a riddle through the parable of the Good Samaritan (Luke 10:30–36). In this parable, Jesus flips the terms of the question: instead of defining or listing the Other(s) for whom we are responsible, Jesus defines when we ourselves are being a neighbor, when we have truly entered into the relationship of neighborliness. Jesus narrates how a man of indeterminate identity was stripped, robbed, beaten, and left by the side of the road, only to have a stream of seemingly righteous people pass him by until he is finally cared for by a Samaritan (a member of a group considered to be of low social standing). Jesus's masterful story keeps our attention on the unknown beaten man, and we expect to find out if this man qualifies as a neighbor for whom the other characters are morally responsible. Instead, as in any good riddle, the terms unexpectedly shift, and we are asked to determine which of the passersby were neighbors (Luke 10:36). And then the terms shift again, as "neighbor" is turned from a geographical or ontological designation to an action, that of showing mercy. We are being a neighbor—that is, following God's commandment—when we show mercy to others (Luke 10:37).

Climate change offers us an opportunity to practice this mercy by which we know ourselves to be neighbors. And given the planetary reach of the crisis, we are called to radically expand our understanding of the neighbor we are commanded to love. We need to broaden the spatial idea of the neighbor, moving from a notion of the neighbor as someone in close physical proximity to us (as inheres in the English word's root "nigh," an etymological inheritance from Germanic languages and still present in the modern German *Nachbar*), to a more expansive notion of those with a shared present.[12] The long-term physics of climate change, with both its causes and its repercussions extending across generations, also challenges us to think of the neighbor in wider temporal terms. The neighbor to whom we should show mercy can be a person far away in both space and time. The neighbor can be a person of the future. Or, to reverse this idea in the terms of Jesus's parable-riddle of the Good Samaritan, showing mercy to the person of the future makes us into a neighbor, enters us into the human

12. The English word "neighbor" primarily signifies proximity, but can also carry a biblical tradition of universal care. The Oxford English Dictionary Online definition A.1 for "neighbor" as a noun reads, "A person who lives near or next to another; a person who occupies an adjoining or nearby house or dwelling; (more widely) each of a number of people living close to each other, esp. in the same street, village, etc.," whereas A.1.b offers simply: "In echoes of biblical passages teaching responsibility, etc., towards others (such as Matthew 19:19): a fellow human."

relationship that God has called for. This loving, merciful human relationship is also how we enter into right relationship with God.

Thinking about futurity, and the future human beings who are our neighbors, is not, however, a strength of Christian theology. The first Christians expected the imminent return of Christ, the Parousia, and so thinking about an ethical responsibility towards the human beings who would live ten, a hundred, or a thousand years in the future was theologically moot, even absurd. Medieval theologians, whose belief in purgatory produced a cultural Christianity that emphasized profound ties of obligation between those living in the present and those who lived in the past, were also not focused on the human future. The Reformation led to a resurgence of interest in the Parousia, which again negated contemplation of future generations. Enlightenment philosophies, while often interested in history, were not focused on future human beings, either. And while the twentieth century saw another theological swing towards emphasizing the eschaton (the end times of human history), this drew attention not towards future earthly human beings but towards the horizon of the ultimate divine promise. Modern ethical concerns have been focused on social justice for those in the present or on the immediate fate of the unborn, but not on social justice for generations of the unborn.

All of these discussions since the time of Jesus have taken place in particular cultural contexts, and it is important to consider that until about the turn of the twentieth century the dominant mode of Western narrative was retrospective—stories looked backwards into time, towards a mythical golden age, towards ancient gods, towards the fabled age of chivalry. With few exceptions, for the first 1,900 years after Christ the human imagination and storytelling looked to the past or the present, but rarely to the future. Today, of course, science fiction is a dominant genre, giving us rich pathways for imagining and even empathizing with future human beings. Literature has often led the way in formulating new ethical considerations, and here our theological and philosophical imaginations are perhaps still catching up with the future-oriented turn of contemporary culture.[13]

In the realm of philosophy, it was only in the last quarter of the twentieth century that scholars started to formulate an ethics that took the people of the future into account. In a foundational essay collection called

---

13. A prominent exception to this is the Left Behind book series about the "Rapture" by Tim Lahaye and Jerry B. Jenkins, which have marked popular evangelical imaginings of the future, often to the detriment of the present.

Obligations to Future Generations, the editors observe, "Until the development of nuclear weapons it seemed obvious to almost everyone that mankind had an almost unlimited future. The questions of whether there could be an obligation to make sacrifices to ensure the continued existence of the human race didn't arise because no one thought that any such sacrifice would ever be necessary."[14] The idea that "everyone" thought there would be an unlimited human future occludes strands of millenarian Christianity, but elsewhere in the volume we find the problem of futurity presented in terms that resonate with the parable of the Good Samaritan and its interrogation of the idea of "neighbor." Gregory Kavka contemplates "whether future people are different, in morally relevant ways, from present people." He asks, "if we think of 'strangers' as those to whom one stands in no such special relationships [of love, friendship, or contractual obligation], are the interests of *future* strangers worthy of equal consideration with those of presently existing strangers?" The answer is yes: "*if* we are obligated to make sacrifices for needy present strangers, then we are also obligated to sacrifice for future generations."[15] Ethically speaking, difference in temporal location does not justify different ethical treatment.[16]

This idea that the present generation has responsibilities to future generations has become an established tenet of environmental ethics.[17] Avner de-Shalit's argument for the philosophical concept of a transgenerational community again echoes the question—"Who is my neighbor?"—at the heart of the parable of the Good Samaritan: "We must therefore ask who the members of our community are. Is our neighbour a member? Are all the citizens of our country members of our community, or perhaps only those whom we like? Or would all the people in the world be members? Were our grandparents members? Suppose you are a member of my community, will your great-grandchildren also be?"[18] As with Jesus's parable, the answer to the question of community membership is expansive. De-Shalit provides "the very definition of what a community is: it is more than simply an incidental, random aggregation of members, or a passing episode, or a functional gathering for one purpose. It is therefore quite obvious

14. Sikora and Barry, "Introduction," vii.

15. Kavka, "Futurity Problem," 187.

16. Kavka, "Futurity Problem," 188, 201.

17. Desjardins, *Environmental Ethics*, 77. See also the essays in Dobson, ed., *Fairness and Futurity*.

18. De-Shalit, *Why Posterity Matters*, 19.

that, although some people die and others are born, the same community remains, and the essence of that community is continuity and succession."[19]

This relatively new philosophical focus on the future has led to calls for a revolution in ethics, for a transformation of the basic premises of ethical decision-making that accounts for future people in ways that classical ethics did not. Hans Jonas has defined the problem in ways that serve as a bedrock for my project, so I will quote him at length. In traditional ethics, he writes,

> The good and evil about which action had to care lay close to the act, . . . and were not matters for remote planning. This proximity of ends pertained to time as well as space. The effective range of action was small, the time span of foresight, goal-setting, and accountability was short, control of circumstances limited. Proper conduct had its immediate criteria and almost immediate consummation. The long run of consequences beyond was left to chance, fate, or providence. Ethics accordingly was of the here and now, of occasions as they arise between men, of the recurrent, typical situations of private and public life. The good man was the one who met these contingencies with virtue and wisdom, cultivating these powers in himself, and for the rest resigning himself to the unknown.
>
> All enjoinders and maxims of traditional ethics . . . show this confinement to the immediate setting of the action. "Love thy neighbor as thyself"; "Do unto others as you would have them to do unto you" . . . Note that in all these maxims the agent and the "other" of his action are sharers of a common present. It is those who are alive now and in some relationship with me who have a claim on my conduct as it affects them by deed or omission. The ethical universe is composed of contemporaries, and its horizon to the future is confined by the foreseeable span of their lives. Similarly confined is its horizon of place, within which the agent and the other meet as neighbor, friend, or foe, as superior and subordinate, weaker and stronger, and in all the other roles in which humans interact with one another. To this proximate range of action all morality was geared.[20]

Traditional ethics, then, were individualized, localized, and presentist. Jonas contends that the technological change of the modern world necessitates a

19. De-Shalit, *Why Posterity Matters*, 21.
20. Jonas, *Imperative of Responsibility*, 4–5.

9

new understanding of ethics, one which "adds a *time* horizon to the moral calculus"[21]:

> All this [the condition of traditional ethics] has decisively changed. Modern technology has introduced actions of such novel scale, objects, and consequences that the framework of former ethics can no longer contain them . . . To be sure, the old prescriptions of the "neighbor" ethics—of justice, charity, honesty, and so on— still hold in their intimate immediacy for the nearest, day-by-day sphere of human interaction. But this sphere is overshadowed by a growing realm of collective action where doer, deed, and effect are no longer the same as they were in the proximate sphere, and which by the enormity of its power forced upon ethics a new dimension of responsibility never dreamed of before.[22]

Jonas then distills this transformation into a new ethical imperative: "Act so that the effects of your action are compatible with the permanence of genuine human life" (or, expressed in the negative, "Act so that the effects of your action are not destructive to the future possibility of such life").[23] In this "your," individual and collective responsibility converge, and the radius of ethical impact expands.[24] This imperative asserts "that we do not have the right to choose, or even risk, nonexistence for future generations on account of a better life for the present one."[25]

ETHICS IN ACTION

In recent years, this ethical imperative to care for future generations has been entering Christian discourse. As Christians of many stripes have increasingly recognized the ethical obligation to respond to climate change, a number of books have considered the environmental crisis as an occasion for engagement at the level of both the church and individual spirituality. In reading these books, one finds many references to the people of the future.

21. Jonas, *Imperative of Responsibility*, 12, original italics.

22. Jonas, *Imperative of Responsibility*, 12.

23. Jonas, *Imperative of Responsibility*, 11.

24. This convergence of the individual and the collective here is a happy grammatical consequence of the English "your" signifying both singular and plural. In the original German text, however, the "you" is a second person singular (*deiner*); Jonas's imperative is quoted in the German in Dalferth, *Umsonst*, 140 n. 18.

25. Jonas, *Imperative of Responsibility*, 11.

One proclaims: "God and history teach us that we must love those least able to defend themselves, which includes the unborn generations of all species"; "We can help ensure that unborn generations will arrive on a healthy planet that needs and welcomes them"; "We must actively help those least able to speak for themselves—including unborn generations"; "God loves all of our neighbors—the ones we don't like, the ones we don't know, and the ones yet to be born."[26] And another asserts: "We must recognize that future generations are no less our neighbors than those who live next door today"; we need "to embrace the universal principle of the Golden Rule by expanding it to recognize future generations as our neighbors: Golden Rule 2.0"; "A repurposed church that explicitly values continuity of creation could declare our moral interdependence with our billions of neighbors the world over as well as our countless yet-to-be-born neighbors."[27]

Rev. Jim Antal maintains that the new role of the church is to initiate "a healing and transformational process" in the context of conversations about climate damage, "and when we do, our generation's obligation to those unborn will motivate us to make the changes that science says we must."[28] While I absolutely agree that institutional churches should provide the structure and impetus for climate action, and while the references to the unborn in religious discussions of climate change are crucial, I feel that we need a more developed and a more complex way of thinking about the people of the future. That we have an obligation to the unborn is a premise of much religious writing about climate change, but beyond this general observation we lack in-depth thought about those future generations in ethical and spiritual terms. As Antal contends, we need motivation to change our fossil-fuel based ways (and fortunately there are rapid advances in renewable energy that increasingly make this possible). But the usual gesture towards the unborn risks becoming a platitude rather than an ethical imperative, too quickly invoked and forgotten to provide deep motivation. When it comes to weighing the future consequences of our actions, modern Americans can be notoriously limited; it is hard to get people to save for their own retirement, let alone focus on saving future generations. Appealing to people's reason or emotions as a motivator for climate action has also proven frustratingly futile, and even counterproductive. The psychologist and economist Per Espen Stoknes, in a chapter

26. Sleeth, *Serve God*, 23, 33, 40, 66.

27. Antal, *Climate Church*, 58, 61, 74.

28. Antal, *Climate Church*, 77.

titled "The Psychological Climate Paradox," observes that "inadvertently, conventional climate communications have triggered more distancing, not increased concern and priority."[29] The more strident and desperate the scientific warnings about climate change, the more people have put their hands over their ears and bought bigger SUVs.

This book seeks to address the lack of focused Christian theological attention on the people of the future with the aim of gently folding those future people into our spiritual lives. The clarion call to climate action has spurred some admirable Christian soldiers into righteous warfare—Antal's book is chock full of accounts of his valiant civil disobedience and arrests. For most of us, though, that is not our path. As a mother, teacher, and mortgage-payer, I don't have the time, means, or stomach for jailtime. But as a Christian, I can clearly hear the message of the environmental and human catastrophe rolling towards us, and can recognize this as a form of spiritual crisis as well. Stoknes writes of the impediments to thinking about climate through five alliterative terms: Distance, Doom, Dissonance, Denial, and iDentity. And he suggests five alliterative terms for communication that will counteract these: Social, Supportive, Simple, Story-Based, and Signals.[30] People of faith can add another S-word to this list: Spiritual. For some, the spiritual motivation to take up arms against climate change will make them firebrands. For others, reflective integration of the people of the future into our horizon of compassion and caritas will be the motivation for right actions. On the level of personal action, this can become the practice of daily consumer choice with an eye towards reducing our dependence on fossil fuels, or giving charitably to the future, such as purchasing carbon offsets to mitigate the environmental impact of one's travel or daily commute.[31] On the level of collective political action, this can become steady coordinated communication with our government representatives, demanding that they support legislation advancing the expansion of renewable energy and reducing pollution.[32] On the level of prayer, this can be

---

29. Stoknes, *What We Think About*, 3.

30. Stoknes, *What We Think About*, chapters 7 and 8.

31. Sites such as Cool Effect (https://www.cooleffect.org/) offer ways to offset your carbon footprint, using donations to help ranchers in the U.S. protect grassland, schools in Malawi to have cookstoves that reduce deforestation, and families in Costa Rica to build wind turbines (per the projects featured on their website January 25, 2019).

32. There are many organizations promoting such action. My own personal involvement is with the non-partisan Citizens' Climate Lobby, https://citizensclimatelobby.org/.

holding future generations in our personal and communal devotions, just as we remember the departed.

The fight against climate change has been described in terms of the abolitionist movement of the nineteenth century.[33] It is a long process, not a quick fix; it will require re-adjusting of fundamental economics; it will need some outspoken leaders and a dedicated army of humble people doing the right thing. It will need faith in a long moral arc bending towards justice (to borrow a borrowed quote from Martin Luther King Jr.), the recognition that we are the fallible movers of this arc, and the realization that this arc will really be more of a zig-zag. It will require actions, and a fundamental trust in the human ability to adapt and change. It will require faith in God.

ABOUT THIS BOOK AND ITS AUTHOR

Since thinking about the people of the future has not been within the purview of traditional ethics or Christian theology, there is not a very deep well of resources to draw from, and what is offered here is of course very small in scale given the magnitude of the topic. But my hope is that these tentative reflections can prompt deeper consideration of future generations in our spiritual, ecclesiastical, and communal lives. The book is grounded in the following precepts: Climate change is not just a scientific, political or economic problem, but an ethical one. The ethical concern is not just one of care for the natural world as God's creation (although that is crucial), but one of care for other human beings—in theological terms, the neighbor. The neighbor is not only the person of the present, but also the person of the future. But while we can clearly make the ethical case for taking care of our contemporary global neighbor, whose face we can see, it is more difficult to care for unborn persons in an unknown future world. It is these future persons—our inheritors, as I name them in chapter 1—who will be suffering the most from our failure to stave off climate change. For most of us, vague platitudes about our grandchildren are not enough to motivate an ethical change of course. What is required is a creative and thoughtful way to place ourselves more deeply into the ethical problem of climate change and the future.

33. This point was powerfully and inspirationally made by Pennsylvania State Representative Christopher M. Rabb in a speech he made at the teach-in and town hall "Rise for Climate, Jobs, and Justice" held at the First Unitarian Church in Philadelphia on September 8, 2018.

The reflective essays collected here are my own beginning of that process. In the first chapter, I outline the problem of climate change as both an environmental and an ethical problem, considering in particular how we conceptualize the people of the future. In the second chapter, I examine an aspect of Christianity that has historically limited or even obstructed theological speculation about our inheritors—the "embarrassment" (to use a term that recurs throughout various strands of biblical scholarship) of the Parousia, Christ's Second Coming. In chapter three, I turn to the representation of generations within biblical contexts (specifically First Kings and the New Testament), reflecting on how we might fold the people of the future into our celebration of the eucharistic liturgy. In chapter four, I contemplate the paradoxical dynamics of the word "individual," which signifies both distinct persons and a larger collective; this paradox of the person who is at once separate and inseparable from others is at the heart of a Christian trinitarianism that can model our relationship with humanity across time. And in chapter five, I explore how the resurrection and its story of forgiveness can serve as a hopeful narrative of futurity, replacing defeatist stories (especially the Greek myth of Icarus) as a way to move forward in our work to combat climate change.

Throughout this book, I write in the first person, with both the singular pronoun of "I" and the plural pronoun of "we." A quick word about the "I," since my "I" is naturally writing from a particular theological and socioeconomic point of view. I am Protestant, although eclectically so. While both of my parents' families were Methodist, I was raised in a fairly liberal, Bible-focused Congregational church near Wheaton, Illinois, the home of Wheaton College (the alma mater of Billy Graham) and a vibrant evangelical community. In Wheaton I attended a Catholic high school in the Franciscan tradition, which left a strong spiritual mark. So my early years were a mish-mash—or, put more artistically, a quilt—of various Christian perspectives. In college I followed secular tradition and dropped out of organized religion, although in graduate school religious literature and history, specifically that of sixteenth- and seventeenth-century England, became my field of study. In my mid-thirties, with new baby in arms, I returned to formal religion, this time through the Episcopal Church, which has become my spiritual home.

Anglican theology, a touchstone of the American Episcopal faith, is marked by a *via media* (a middle way) between Protestantism and Catholicism, and often by a tendency to rest in the spiritual space of paradox. I have come to recognize how the tendencies of Anglican theology also inform my climate politics. If I were to label the political party to which I want to belong, it would be "pragmatist," and in a *via media* way I look for climate solutions that might come from either the right or the left of the political spectrum, so long as they would be effective in saving the planet and humanity. I recognize that I myself fall into paradoxes on climate action—I drive my car to climate activist meetings, for instance. I believe that we need to move quickly on addressing climate change, but while I am steadily evangelical on this mission, I don't think we are helped by absolutes. I am inspired by the enthusiasm of younger climate activists, who have prompted me to think seriously about capitalism (as have the writings of Michael S. Northcott, whom I discuss in chapter one, and Kathryn Tanner, discussed in chapter four). It is not clear to me, though, that we can change the whole socio-political-economic system in time to mitigate climate change, and so while I think many of the critiques of neoliberal capitalism are apt, at this historical moment we need to consider how to address climate from within the system, even as we contemplate a different economic future. Someone once noted that nineteenth-century abolitionist women did their work wearing cotton dresses: it is a striking and powerful image when you stop to think about it.[34] The women were striving to change an unjust economic system even as they were clothed in the very products of that system, given no available alternatives. For some time to come, we will be in the same strange situation, flying to conferences on climate change, or writing books on climate change in a home warmed by natural gas. This is the historical moment in which I find myself.

In terms of my point of view I am also female, white, middle-aged and (upper) middle-class. I am American; while I am keenly aware that other cultures have their own configurations of religiosity and climate challenges and solutions, I am writing from the place I know. And for what it is worth, I am an English professor, which I have been told shows in my mode of biblical reading—hopefully in a good way.

If that is the me, the "we" of this book also requires explanation. In terms of religious identity, I have presumed that readers drawn to this book are primarily Christian, so the "we" often refers to Christians writ large. That said, this is not an exclusionary "we": if a reader has come to this book from a different religious tradition, or is even part of the growing number

---

34. With apologies to the person who first made this comparison, I haven't been able to locate where I read or heard this point.

of "nones" (people with no religious affiliation), welcome. Politically, here's what I hope the "we" is not: a differentiation of an "us" and a "them." One of the tragic and dysfunctional effects of the climate culture wars is that even the idea of our children's future has been politicized—and polarized, as different populations fight to protect a long-term or a short-term future, the projected health of global ecosystems or the immediacy of a secure job. With a sense of existential urgency, some people are doing all that they can to move swiftly towards a future that is free from fossil fuels, and others are doing all that they can to maintain a fossil fuel industry that provides employment for their family. A juxtaposition of same-day newspaper articles illustrates this: one article tells about a children's climate lawsuit that argues for a "constitutional right to a stable and safe climate" and demands a "plan to phase out fossil fuel emissions"; another reports on an oil and gas workers' rally to oppose fracking restrictions in Colorado, where one man attends with his four small children (Parker, Pepper, Peyton, and Preston) and another with his twins (Everleigh Halli and Adler Burton, collectively named after Halliburton, a major company in the oil industry).[35] Each of these groups of people, those involved with the lawsuit and those attending the rally, seeks their children's safety and happiness, and each sees the other as a serious threat to their hopes for their children's future. It is difficult to find common ground here: a healthy planet and a robust fossil fuel industry cannot coexist. There is going to be cost in the future, in terms of either a changing climate (and the attendant social, environmental, and economic damage) or changing industries (and the attendant shift in jobs and communities). But while fossil fuels and climate are inimical, we ourselves do not need to be tribal and hostile. The future, one way or another, is going to be different, and I hope to find a "we" that is generous, expansive, and inclusive.

It is also important to acknowledge that responsible environmental stewardship and the mitigation of climate change requires making choices that move us away from a carbon-intensive lifestyle, but choice is often a privilege. People who live in food deserts, or who cannot properly maintain their vehicles or afford new ones, or who have no leverage over landlords that do not insulate and responsibly heat tenants' apartments, all contribute to our carbon-intensive world, but not usually through conscious acts of choice. The ability to choose confers moral responsibility. The "we" of this book, then, is anyone who is in a position to make carbon choices,

35. Schwartz, "Young People Are Suing"; Turkewitz and Krauss, "In Colorado, a Bitter Battle."

whether small (adjusting the home thermostat) or large (enacting political legislation or corporate policies that move us away from a carbon-based economy). We need to recognize that care for the poor is also care for our ecosystem: one study in Toronto revealed that less than 25 percent of cars caused 100 percent of the black carbon automotive emissions.[36] Environmentally speaking, someone's old clunker automobile is everyone's problem. Part of moving towards a better environmental future is equipping everyone with better carbon options (more efficient public transit and renewable energy, for instance) and better living wages.

Fundamentally, on an ethical, spiritual, and pragmatic level, these two first person pronouns—the "I" and the "we"—are not distinct. As I will discuss in chapter four, one of the most fascinating words in the English language is "individual," which means both a unique and separate person, and an undividable, indivisible whole. The word thus presents a paradox, containing both the personal and the collective. These two seemingly oppositional definitions rest together: the "me" and the "we" are mutually intertwined. In what follows, I sometimes choose to imagine climate scenarios from a first-person singular perspective, and sometimes from a first-person plural point of view. The personal and the collective are not only ethically interlinked, but environmentally; this interconnection is an inherent element of the dynamics of anthropogenic climate change, since personal and collective carbon emissions cannot be distinguished in terms of the damage they do. Consequently, the solution to climate change also entails both personal and collective action.

There are plenty of reasons to lose faith in our ability to mitigate the effects of climate change. And we have biblical warnings about how human actions have environmental impacts. In Micah, we read that "the earth will be desolate / because of its inhabitants, / for the fruit of their doings" (Mic 7:13). But that story is not yet finished. The psalmist tells us that "those blessed by the LORD shall inherit the land" (Ps 37:22). We are the agents of that change who can pass along a healthier land to our manifold heirs.

36. Wang et al., "Plume-based analysis," 3263.

1

Changing

An Environment in Peril and a Christian Response

You do not even know what tomorrow will bring. What is your life? For you are a mist that appears for a little while and then vanishes. Instead you ought to say, "If the Lord wishes, we will live and do this or that." . . . Anyone, then, who knows the right thing to do and fails to do it, commits sin.

—JAMES 4:14–17

AN ENVIRONMENTAL PROBLEM

How does the lay person, the non-scientist, imagine and respond to global climate change? The factors that are creating the problem can be described in some very big words—hydrochlorofluorocarbons, Atlantic multidecadal oscillation, Holocene climatic optimum—but those don't really help me out. Traditionally, the image used to describe the physics of climate change has been that of a greenhouse: carbon dioxide (CO_2) emissions collect in the earth's atmosphere and act like the glass roof of a greenhouse, so that the sun's warmth can come into the greenhouse but cannot leave in the same way, thereby trapping heat.[1] As an English professor who spends her

1. For an excellent animated graphic that quickly demonstrates the greenhouse effect,

time working with metaphor, I think the word choice of "greenhouse" is a bit dumb. Maybe the image of a greenhouse does a decent job of presenting a process, but it does a lousy job of presenting a problem. My connotations of "house" are positive; we can think of the coziness of "dwelling," of "hearth," of "home." And green—don't we all want to be green these days? "Green" has entered the advertising mainstream as a vague ecological positive, a word that signals good (consumer) choices. "Green house" doesn't sound that bad to me, in fact it sounds both appealingly domestic and environmentally responsible. When I think of actual greenhouses, I think of luscious, wondrous places in which deeply fragrant and exotic plants like orange trees and orchids luxuriate.[2] Or I think of practical places where healthy food is grown, or where beautiful garden plants are cultivated and nurtured. I wouldn't mind being in a greenhouse right now.

Many wise people have recently tuned in to the fact that *how* we talk about climate change is part of the problem of climate change. Word choice matters, and fuzzy knowledge of some of the basic working terms impedes comprehension of the climate issue. People confuse weather and climate, and carbon monoxide and carbon dioxide, and often vaguely remember from school that CO_2 is involved with trees. They conflate the climate crisis and the "hole" in the ozone layer (phenomena that are in fact more the inverse of each other, the former caused by the atmospheric build-up of gasses, the latter resulting from the atmospheric depletion of a gas).[3] If the

---

see Australian Government, Department of Environment and Energy, "Greenhouse Effect."

2. In fact, the first use of the term "greenhouse effect" in *The Geological Magazine* (1867) directly invokes this imagery: "The atmosphere . . . would permit the solar heat to pass through . . . but would prevent its escape by radiation after it had once heated the surface of the earth, and would thus immensely augment the temperature . . . precisely as if we had covered the whole earth with an immense dome of glass,—had transformed it into a great Orchid-house" (*Oxford English Dictionary*, def. 3, first historical example). The coinage of the term "greenhouse effect" coincided with a Victorian cultural and architectural vogue for greenhouses; London's great Crystal Palace for instance, was built in 1851. Svante Arrhenius, the nineteenth-century scientist who discovered the connection of CO_2 emissions and a rising temperature of the earth, believed that the burning of coal would lead to a more equitable global planet, as colder climes would have more abundant crops (see quotes collated on Wikipedia, "Svante Arrhenius"). The favorable visualization of the earth as "a great Orchid-house" unfortunately did not predict the vast flooding of Midwestern farmland or the extensive drought experienced by Australian ranchers.

3. For instance, one participant in a recent poll "acknowledged uncertainty about the cause of climate change. 'The way I always understood it, I thought global warming

person on the street cannot explain the problem of climate change, they are unlikely to respond to it. Recently I attended a workshop on climate communications led by the National Network for Ocean and Climate Change Interpretation (NNOCCI). There was a lot of helpful information on how to explain the problem, such as distinguishing between *regular* CO_2 emissions—everyone is doing it! humans and animals emit CO_2 when we breathe—and *rampant* CO_2 emissions, the overload we are pumping into the atmosphere through burning fossil fuels. CO_2, on its own, is an integral part of the wondrous, intricate design of the ecosystem we inhabit. But we are upsetting this system through our CO_2 excesses.

As part of our communication training exercises, we learned that the greenhouse model wasn't working well to convey the problem. (Evidently I am not the only one who feels warmly about greenhouses.) We were taught that a better way to express the problem is to think of the atmospheric build-up of CO_2 as forming a heat-trapping blanket around the earth.[4] The planet is always blanketed by our atmosphere, of course, otherwise we would be like Mars, where temperatures can range from a comfortable 68°F in the daytime to -119°F at night.[5] The problem occurs when excess CO_2 makes earth's blanket thicker, turning it from a gauzy shawl to a heavy piece of felt through which heat can't escape. This model can be easily demonstrated using people: wrap someone in a thin piece of fabric on a sunny day and they are fine, but wrap them in a thick blanket for a few minutes and their body temperature quickly rises.

The static metaphors of the greenhouse and the heat-trapping blanket might help me to better understand environmental physics, but these images still don't quite explain why planetary warming is a problem and what we are supposed to do about it. So I have developed my own highly

---

had more to do with the fact that we've broken down so much of our protective layer,' she said, referring to the destruction of atmospheric ozone by chlorofluorocarbon gases, which have been banned since 1986 under an international treaty"; Guskin, "Americans Broadly Accept Climate Science." This treaty, the Montreal Protocol, is a model of international environmental cooperation that is now having real effect. According to NASA, "The ozone hole over Antarctica is expected to gradually become less severe as chlorofluorocarbons— banned chlorine-containing synthetic compounds that were once frequently used as coolants—continue to decline. Scientists expect the Antarctic ozone to recover back to the 1980 level around 2070"; NASA, "2019 Ozone Hole."

4. For a succinct account of the physics of the blanket model, see the American Chemical Society's website, "A Greenhouse Effect Analogy."

5. These are the extreme July temperatures for 2012–2015; see the temperature chart on Wikipedia, "Climate of Mars."

unscientific model that I think does a better job of imagining the problem, if not accurately depicting the process, of climate change. Picture a bathtub: there is a faucet where the water comes into the tub, and a drain where the water can leave the tub. Let's pretend that you are not actually going to take a bath, but for some reason you leave the water running. As long as there is an equal amount of water simultaneously pouring into and draining out of the tub, you are fine. But let's say that you increase the amount of water flowing out of the tap. Now there is more water going into the tub than can leave at the same rate. At first, you do not have real problem—the excess water simply accumulates in the tub. But let's say the water is left running for hours. Now the tub is really filling up with water. And let's say that the water is not just left running, but that you actually keep turning the faucet so that the water is gushing out more and more. Before too long, the tub is full. There is still water draining out of the bottom, of course, but not fast enough to manage the inflow. The water starts to dribble over the side of the tub. Then it starts to pour over the edge, and onto the bathroom floor. You run around and grab towels to try to mop up the excess. But soon you have used up all of your towels, and it's not solving the problem. The water continues to flow for days, months, years. It starts to run down the staircase. It gets into your living room, and destroys old family photo albums, and books, and your children's artwork. It gets into your kitchen and ruins your food. It gets into the plaster and the drywall and the rafters . . .

As a scientific analogy, my bathtub probably doesn't, well, hold water. But as a way to picture the problem on a pragmatic level, I think it does a good job.[6] The idea of standing in a greenhouse, or of being wrapped in a blanket, doesn't really convey the dangerous environmental consequences of climate change. From what I can make out from my lay person's reading, we are at the historic moment when the water is starting to trickle over the edge of the tub—or perhaps it is already seeping into the walls. I think

6. After I had developed my bathtub analogy, I found that Christiana Figueres, a Costa Rican diplomat, similarly described the process of climate change as caused by "pouring dirty, poisonous sludge into a bathtub with a partially opened drain"; summarized in Robinson, *Climate Justice*, 5. My friend (and friendly geologist) Dr. Matthias Ohr has suggested to me in conversation that another way to think of things is that there are really two bathtubs, that of the slow carbon cycle (which moves carbon through the earth and rocks) and that of the fast carbon cycle (which moves carbon through life-forms on earth). In the process of extracting and burning fossil fuels, we have effectively switched the faucet from the large tub (slow cycle) to the small tub (fast cycle), which isn't prepared to handle the influx. For a clear description of the two cycles, see Riebeek, "The Carbon Cycle."

my analogy is obvious, but just to spell it out: The water flowing out of the faucet represents CO_2 emissions. Again, CO_2 is a normal and necessary part of our ecosystem; we need CO_2 flowing into the atmosphere as part of a healthy, balanced world. But starting with the Industrial Revolution, we have been pouring more CO_2 into the air than the system can handle. For a long time this wasn't much of a problem, as the planet's carbon "sinks" (most notably the ocean) took in the excess and the absolute quantity of emissions remained relatively small.[7] But as we have continued to increase our emissions in the last half century, a larger portion of the excess carbon increasingly cannot be handled through the earth's normal environmental processes. And so it builds up, causing the changes in climate that threaten our homes, our food supply, our health, our cultural heritage, and our children's future. The bathtub analogy not only captures a sense of the problem and the consequences of pouring carbon emissions into the atmosphere, but also indicates that action is required on our part. Picturing myself standing in the rich sensory space of a greenhouse, marveling at the vegetation around me, or picturing myself standing, weirdly, in a big blanket on a summer's day, I can't imagine what I am supposed to *do*. But as a homeowner, I can sense the urgent need to take action when my tub is about to overflow.

Of course, the obvious thing to do when your bathtub is about to overflow is to *turn off the water*. And this is where things get difficult. Because while you can imagine the natural environment as the house you live in with your own family, none of us has individual agency over the collective flow of CO_2 emissions. While there are specific actions and decisions you can take to reduce your own burning of carbon—you can improve the insulation of your house, buy more local foods and reduce food waste, get more efficient lightbulbs, and sign up for a utility company that uses renewable sources of energy, for instance—the burning of fossil fuels is such an integral part of our daily lives that it is nearly impossible, on an individual level, to de-carbonize your life. Think of the carbon-heavy journey taken by your toothpaste, or your dogfood, or your cellphone before it arrived in your hand; you might give up eating out-of-season fruits that require lots of transportation to get to your plate, but other things seem non-negotiable. Or you might change something in your transportation life—perhaps you bike to work one day per week, or carpool more, or buy an electric vehicle—but what about flying with the kids to visit their grandparents? We are

7. Hayhoe and Farley, *Climate for Change*, 78.

often facing choices that don't seem like choices, and personal (non)choices that happen within a de-personalized system. Turning off the faucet of our communal carbon emissions is going to require both individual and collective actions. It is not going to be easy or quick.

Another problem with turning off the faucet is that there is no clear answer as to who turned it on in the first place. Do we hold "The Industrial Revolution" accountable? Do we hold other countries accountable? Do we hold oil and gas companies liable, even as we were the ones consuming the fossil fuels they produced? And what would holding others accountable even look like? If, say, a plastic container is thrown in a recycling bin in the U.S., shipped to China and transformed in a carbon-intensive factory into a children's toy, which is then shipped back to the U.S. and transported across the country by a diesel-fueled vehicle to a store or warehouse, and then is purchased by someone who drove to the mall or who ordered the product and had it delivered directly to their door, whose carbon footprint is it? And then there is the question of where the original plastic container was produced and how it was transported in the first place, and before that how the petroleum or natural gas products for making the plastic got to the container production plant. Soon the "carbon footprint" comes to resemble not a clear set of tracks on a pristine stretch of white sand, but rather the overlapping traces of thousands of beachgoers after a summer day at the New Jersey shore. Consequently, the carbon culpability is not individualized, but distributed. And therefore it quickly becomes clear where we should point the accusing finger. I have spoken as a homeowner, but now I will also speak as a parent: "Not Me" has been responsible for many mishaps in our home. Part of the problem of addressing global climate change is that it was also created by Not Me.

Another problem with addressing the very problem itself is that the person it will affect is also not me—or probably not me, or hopefully not me. Increasingly, we see images of climate-related damage where the effects of the overflowing bathtub—or, more specifically, the effects of storm systems that both hold more water and move more slowly due to warmer ocean temperatures—are painfully clear (see Figure 1). But until fairly recently, it seemed as if the consequences of climate change would happen in a hazy, distant future, a future in which people would have invented technology that would somehow take care of our carbon mess.[8] So the problem

---

8. As recently as the 1980s, many scientists thought that the effects of climate change would be felt in centuries or millennia; the effects that are already manifest today would

was a future problem for future people and might not even turn out to be a problem because of their future gizmos. We are already witnessing the consequences of a changing world climate—fires and floods, droughts and famine—but on the whole, for many people in North America, this still seems like it is going to be someone else's problem. Not mine.

Figure 1. A home flooded by Tropical Storm Erin in Kingfisher, Oklahoma, 2007; FEMA Photo Library/Marvin Nauman.

A MORAL PROBLEM

Climate change, then, poses not only a global environmental issue on a scale that has never been faced by human beings, but it also presents a unique set of ethical dilemmas in terms of assessing culpability and response. The causes of climate change lie in the past and present, and the consequences affect the present and future. At this point in time, we are both the victims and the perpetrators of environmental damage. We have both inherited a mess and are contributing to a mess; the situation both is and is not our fault. It is both our responsibility and not our responsibility. There is not a terribly clear ethical path forward.

---

have been considered worst case scenarios and dismissed as alarmist in the 1990s. Linden, "How Scientists Got Climate Change So Wrong."

To illustrate the ethical problem on a non-cataclysmic level, we can look at nocturnal temperatures in the city where I live, Philadelphia. Nighttime temperatures in my city have been getting increasingly hotter; in 2017, there were thirty more nights per year with a nocturnal temperature above 65°F as compared to averages in the 1970s.[9] While Philadelphia daytime temperatures are also rising (between 1950 and 1999 there were an average of three days per year that topped 95°F, while a report from the city's Office of Sustainability forecasts that we could see as many as fifty-two 95°F+ days per year by century's end[10]), "the biggest change is happening at nighttime, when temperatures are cooling off less. One big reason is the increased moisture in the atmosphere. The more humid the air, the less the temperature can fall."[11] Muggy nights in Philly are thus part of the same phenomenon that contributes to the massive hurricanes flooding the east coast. I hate sleeping in a warm, humid bedroom. So what do I do on these brutal summer nights? Naturally, I crank up the air conditioning. In so doing, I end up contributing to the problem both by generating more urban heat through my machine, and also by sucking up energy from a coal- or gas-burning power plant. Did I cause this problem of hotter nights? No. And yes. Generations before me poured their CO_2 emissions into the atmosphere: not my fault. But even though I now know the consequences of my own CO_2 emissions—the bathtub overfloweth—I continue to engage in actions that I know are making the summer nights hotter, the seas higher, and my children's future dimmer. I shop online; I buy stuff from IKEA; I drive my kids to school (and to music lessons, and sleepovers, and camp, and college visits, and recitals, and choir rehearsal, and church, and . . . and . . . and . . .). I commute to work; I fly to conferences and to vacation places. By now I have tried to reduce my carbon footprint—I do drive a hybrid, and we did switch to an electricity provider that only has renewable energy sources, so that the A/C at least is no longer running on fossil fuels—but it seems like a drop in the proverbial bucket. Or rather, I recognize that I am still contributing a trickle into the overflowing bathtub. And that, even knowing this, I still do things like buy out-of-season or tropical

---

9. Boren, "Philadelphia Is Getting Hotter." The Northeast has been getting much wetter; see U.S. Global Change Research Program, *The Climate Report*, 117. This report contains sections on climate forecasts for different regions of the U.S., should readers like to read up on the changes in their home region.

10. Avril, "Climate Change is Hurting Philadelphians' Health."

11. Boren, "Philadelphia Is Getting Hotter."

foods (lemons! avocados! grapefruit!), maybe things grown somewhere in a greenhouse that have to be driven or flown to my city.

I confess that I am a climate denier. Not in the usual sense of the term as someone who denies that there is anthropogenic global climate change. Should anyone still need a primer on climate change and its human causes, a great summary description is found in *Caring for Creation: The Evangelical's Guide to Climate Change and a Healthy Environment*, by Mitch Hescox (leader of the Evangelical Environmental Network) and Paul Douglas (a meteorologist).[12] They provide clear descriptions ("Weather is, *Do I need shorts or a jacket?* Climate is the *ratio* of shorts to jackets in your closet") next to sobering facts ("The average American emits about fifteen tons of CO_2 into the air every year just by driving a vehicle. This invisible, heat-trapping blanket of man-made chemicals is the rough equivalent of four Hiroshima-sized atomic bombs' worth of extra heat *every second*. Put another way, that's 400,000 atomic bomb blasts of additional heat energy *every day*").[13] I do not doubt the science—nor the scientists, 97 percent of whom agree that humans are changing the climate[14]—but I am often in denial that I am contributing to the harm being done to the planet, harm that will affect my children and future generations. I am not personally responsible for causing global climate change, and yet I am personally responsible for contributing to global climate change. And the consequences of this environmental change are quickly heading towards the drastic: extinction, famine, transformations of landscape, poverty, suffering, massive population migrations. Ecosystems as well as political systems are endangered.

12. Hescox and Douglas, *Caring for Creation*, chapter 1. For a more extensive account of the science of climate change (and the slow-moving political disaster that has led to government inaction), see Hansen, *Storms of My Grandchildren*.

13. Hescox and Douglas, *Caring for Creation*, 23–24. For the statistic of CO_2 emissions, they cite Thom Patterson, "Will Hacking Nature Protect Us from Climate Change," CNN, October 27, 2015.

14. See the NASA website "Scientific Consensus," a compilation of studies from scientific associations like the American Association for the Advancement of Science, the American Chemical Society, the American Geophysical Union, the American Medical Association, the American Meteorological Society, the American Physical Society, the Geological Society of America, and other government and international bodies. For readers who want a concise account of the science, see World Meteorological Organization, "WMO Greenhouse Gas Bulletin"; this bulletin has a subtitle, "Isotopes confirm the dominant role of fossil fuel combustion in increasing levels of atmospheric carbon dioxide."

My form of denial, then, is that described in the biblical letter of James: "Be doers of the word, and not merely hearers who deceive themselves" (Jas 1:22). I am hearing the word that carbon emissions are damaging creation and the neighbor, and yet I deceive myself into thinking that I am not called to be a doer, someone who takes actions to stop the damage. Again, this is both my individual fault and a broader cultural dilemma. The causes of climate change are so refracted and distributed across systems and societies, and the ethical problem has emerged so rapidly, that I don't have a good way to think of my environmental culpability. But I am starting to confront the fact that my tendency to ignore the CO_2 problem is not distinct from those who continue to profess willful ignorance of climate change: ignoring and ignorance are part of the same continuum.

In his book *A Moral Climate: The Ethics of Global Warming*, Michael S. Northcott, an ethics professor and a priest in the Scottish Episcopal Church, writes, "[L]oss of a sense of place, and of local politics, is deeply implicated in the cultural pattern of denial about the human and ecological consequences of industrial consumerism, of which denial about global warming is perhaps the strongest manifestation."[15] In the day to day, I participate, usually unwittingly and sometimes even gladly, in "the cultural pattern of denial about the human and ecological consequences of industrial consumerism." I am a consumer. I don't know how to be otherwise. Moreover, the economic ethics overwhelm me: even as the rapacious consumption of consumer goods harms people and planet, it is this very consumption that the global economy depends upon. Were we all to somehow abruptly cease consuming, a ruined economy would lead to human suffering and poverty. All told, I generally find it easier to ignore my harmful actions or deny my own agency or deceive myself that there is nothing to be done.

Part of the problem of tackling the problem of global climate change, then, is that we lack a clear moral compass on the issue, not only a way of deciding right from wrong, but more broadly a way to think about climate change with our conscience—and with our God. I was motivated to write this book because of a couple of powerful moments I experienced at my church. First, in December of 2017 the bishop of my diocese, Bishop Daniel Gutierrez, was making his annual parish visitation and came to a reception after the service. From the podium, he asked if there were questions or concerns from the congregation. I would normally not ask a question in such a forum (in a loud, echoing hall with everyone looking at me), but nobody

15. Northcott, *Moral Climate*, 93.

else raised their hand and in the silence I was moved by the Spirit, as the Quakers might say, to ask a question: "What is our diocese or the national Episcopal Church doing about climate change?" Although I didn't intend for it to be hostile, the question may have been a bit aggressive, firmly putting the responsibility on someone else: essentially, I asked what are YOU doing about climate change. The Bishop responded that he did not know of any actions being taken by the Episcopal Church at the diocesan or national level.[16] And then he came back at me, "What are YOU doing within your church about climate change?" I remember growing red in the face. He then went on to say that he had recently seen a video of a starving polar bear eating a discarded sofa, and he said he would absolutely label that as sin.[17] Shortly thereafter in a different presentation, our priest, the Rev. Dr. Joseph Wolyniak, asked, "What happens if we think of every act of turning the key in the car's ignition as a sin?" Episcopalians are generally big on social justice but a bit squeamish talking about sin, so it was startling to hear the word used twice in a short period of time to talk about climate change in a theological context.

In contemplating these two comments, I have come to see the bishop as talking about Sin (the part of the human condition that alienates us all from God, neighbor, and self) and the priest as talking about sin (our individual thoughts and actions that alienate us personally from God, neighbor, and self). If centuries of spewing CO_2 and other pollutants into the air is big-s Sin (the act of sullying our world in the general interest of profit and consumption, "structural sin" in the terms of Liberation Theology[18]), my own particular choices (say, turning up the thermostat in the winter because I don't want to wear a sweater) are small-s sins. Big-s Sin is the stream of water that flowed into the bathtub before I was even born; it is, we could say, original, the state into which I was born, the collective effect of railroads and factories and highways and airplanes and ships. My own

16. The national Episcopal Church subsequently formally endorsed a form of carbon pricing known as carbon fee and dividend at its 2018 General Convention, and has asked individuals and congregations to sign on to a "Pledge to Care for Creation" (https://episcopalchurch.org/creation-care), among other actions.

17. I haven't found the particular video he was referring to, but a National Geographic video on YouTube, "Starving Polar Bear on Iceless Land," conveys the idea; watch https://www.youtube.com/watch?v=_JhaVNJb3ag.

18. For a discussion of how the economic-environmental ecology of climate change can be seen through the lens of structural sin and liberation theology, see Northcott, *Moral Climate*, 153–56.

small-s sins are the daily actions I do that add to the mix. These small-s sins themselves could be further distinguished as sins of commission (the environmentally harmful choices I make, like driving when I don't feel like taking public transportation) and sins of omission (the manifold environmentally helpful things I don't do, like get my leaky basement windows replaced).

At this point it is perhaps helpful to differentiate my CO_2 emissions in a more subtle way; if all of my emissions are sin, then I can only despair that I am beyond redemption. Northcott turns to the work of Henry Shue to think through the distinction between "livelihood emissions" and "luxury emissions." Livelihood emissions result from the necessary "moderate levels of greenhouse gas consumption [required] to grow and cook food, keep warm, build shelters and make festival." Luxury emissions are those used by the rich "to sustain their unparalleled levels of ownership and consumption of everything from property and cars to foreign holidays and entertainment devices."[19] Although Northcott does not define "rich," he is not talking here about the super wealthy, the 1 percent. In the larger context of the global population, someone with a home, a car, an occasional trip to Florida, and an iPhone is wealthy. Northcott continues, in words that are uncomfortable to someone with a home, a car, an appetite for travel, and an iPhone (that is, me): "Luxury emissions represent moral malfeasance by the rich against the poor both nationally and internationally, since they give rise both to local and global forms of pollution and other kinds of moral harms. . . . Once they know of the damage their emissions are doing, the rich have no excuse for not reining in their emissions, and for not compensating the poor for the damage their emissions are doing."[20] (Maybe even more uncomfortably, we read in the Bible, "For the sun rises with its scorching heat and withers the field; its flower falls, and its beauty perishes. It is the same way with the rich; in the midst of a busy life, they will wither away. . . . But one is tempted by one's own desire, being lured and enticed by it; then, when that desire has conceived, it gives birth to sin, and that sin when it is fully grown, gives birth to death" [Jas 1:11, 14–15].)

Northcott (again, channeling Shue) presents a way of conceptualizing individual CO_2 emissions in a way that I find useful. To say that all fossil fuel burning is categorically bad causes me to hopelessly throw up my

19. Northcott, *Moral Climate*, 56. Northcott cites Henry Shue, "Subsistence Emissions and Luxury Emissions," *Law and Policy* 15 (1993) 39–59.

20. Northcott, *Moral Climate*, 57.

hands, but separating out different types of fossil fuel use gives me a way to think more constructively about how I am using them. There are people who urge a rapid switch to a zero emissions lifestyle, but how am I to do that, having been born into the original sin of a carbon-intensive world? Perhaps someday I will be able to bring my carbon footprint to zero. I could replace my current natural gas heating system with geothermal, switch to an electric stove that is powered with electricity produced through renewable means, drive a Tesla that I likewise charge with renewable energy, and perhaps grow my own food. But . . . I don't see that happening soon, and it sounds terribly expensive. I have the life that I have. For the time being, though, I can think of my furnace and oven, my heat and my cooking (and maybe some of my driving?), as "livelihood emissions." But: I am forced to recognize that most of my emissions are in the category of "luxury emissions." Having made that acknowledgment, I then recognize that there is a cost to others of my direct and indirect emissions. Having recognized that cost, as a moral person—and as a Christian—I am called upon to pay that debt. In the words of the Lord's Prayer, I am a debtor.

I instinctively want to avoid this thought. I start to think about loopholes: "What about my Amazon Prime orders? I'm very busy; could those be counted as livelihood emissions? Please?"). And I start to think about how I don't really know how to think about my culpability in contributing to global climate change. While the distinction of livelihood/luxury emissions provides the beginnings of a framework for approaching my own ecological sins, the luxury side of things quickly feels overwhelming. Northcott's account of how global climate change is a consequence of neoliberal economics convinces me but does not comfort me. He writes,

> Global warming is the earth's judgment on the global market empire, and on the heedless consumption it fosters. The neoliberal claim is that the market, combined with technological power, can redeem the peoples of the world from pain and suffering through the autonomous, self-regulating market system. Those who direct the neoliberal project of economic globalization presume that the welfare of people and planet is advanced when the most powerful economic actors—multinational corporations, bankers, investors, engineers—are freed from regulation and taxation to make and sell more "consumer objects" and accumulate more wealth. But in reality this collective project of global wealth accumulation disempowers people in communities of place, and so provokes enormous destruction of the welfare of human communities. . . .

At the same time it presages the greatest ecological collapse in the history of the human species. Global warming, in other words, is the global market empire hitting its biopolitical limits.[21]

On a fundamental level, I reluctantly must confront the Sin/sin of consumerism (the byproduct of capitalism). Again, I was born into this world, so this sin is original, but I know that my lifestyle causes suffering to others. And again, Northcott's haunting words: "Once they know of the damage their emissions are doing, the rich have no excuse for not reining in their emissions." Or, maybe even more cuttingly, "Anyone, then, who knows the right thing to do and fails to do it, commits sin" (Jas 4:14–17).

In the Episcopal Book of Common Prayer, there is a moment in the liturgy of the Holy Eucharist, before we take communion, when the congregation collectively confesses their Sins and sins: "Most merciful God, we confess that we have sinned against you in thought, word, and deed, by what we have done, and by what we have left undone. We have not loved you with our whole heart; we have not loved our neighbors as ourselves. We are truly sorry and we humbly repent."[22] Our carbon emissions harm our neighbors as ourselves. It doesn't seem like a stretch to compare this to the phenomenon of second-hand smoke: once our culture became aware of the health impact of second-hand smoke, the issue was no longer about the individual rights of the smoker but the collective rights of those around us, and spaces that had previously been filled with smoke—offices, college classrooms, homes with children, etc.—became spaces with clearer, purer air. I am old enough to remember smoke-filled restaurants and airplanes (!), and the momentary outrage when smoking was banned in bars (of all places!). But even in the midst of the current acrimonious American culture wars and fervent calls for personal "freedom," I never hear of anyone advocating for the return of cigarettes to restaurants. We somehow managed to come to a communal ethical consensus, and there was a corresponding shift in behaviors. The issue of global climate change is infinitely more complex than banning smoking in public places, but there is a moral analogy here in terms of care for the neighbor. The compassionate ask, though, is a larger one: it is one thing to accept a smoke-free environment for a colleague, a beloved child, or the stranger sitting next to you at a restaurant. But it is more difficult to give up something for a faceless stranger who might be more of an abstraction.

21. Northcott, *Moral Climate*, 7.
22. Episcopal Church, Book of Common Prayer, 360.

For a long time I was feeling guilty about turning down the thermostat on my air conditioning because I would think of the poor polar bears, which I could clearly picture from PBS nature shows, once playing with their young and running on ice, and now starving and drowning—and, horrifically, eating each other—as that ice recedes. (And in truth, my heart breaks for the polar bears, as well as for the thousands of koala bears killed in Australian wildfires that resulted from an extended period of record drought and high temperatures.[23]) But lately I have also been thinking, as I contemplate what I have done and left undone for the environment, "who is my neighbor?" That question is famously posed in the parable of the Good Samaritan (Luke 10:25–37), where Jesus describes a man who is stripped and beaten (a man who thus cannot be clearly identified as belonging to any particular group of people) and the response of others who pass by, crossing the road so as not to have to look at him or engage, absolving themselves of responsibility for the situation. A key point of the parable is that Jesus answers the question about defining the neighbor—defining the human being for whom one is morally responsible—very simply: the neighbor is everyone. The stripped, beaten man was a neighbor to all who passed by him.

On a local level, care for the neighbor embraces all of the inhabitants of my city of Brotherly Love, many of whom are being affected—and will increasingly be affected—by a changing climate. Returning to the phenomenon of rising nocturnal temperatures in Philadelphia, we read: "Warmer nights pose health consequences, because they give bodies less time to recover from heat. That leaves already vulnerable populations—the elderly, people with breathing issues, those without air-conditioning—at higher risk of heat-related illnesses and death. In Philadelphia, 104 people have died from heat-related causes in the last decade."[24] The comfort of my cool home when fueled by carbon has a cost not just for distant polar bears, but for people within walking distance from my front door.

---

23. An estimated 30,000 koalas were killed or injured in the extensive Australian wildfires of 2019/20; Burke, "Video Shows Koalas." A quarter of the koala population of New South Wales is estimated to have died. Beyond the koalas, a staggering number of animals perished in the fires. Kingsley Dixon, a university ecologist and botanist in Perth, is quoted as saying "We will have taken many species that weren't threatened close to extinction, if not to extinction"; Albeck-Ripka, "Koala Mittens." The estimate of one eminent Australian scientist is that 480 million animals (mammals, birds, and reptiles) perished as a result of the season's wildfires; University of Sydney, "Statement."

24. Boren, "Philadelphia Is Getting Hotter."

And there is cost to those across the seas. Katharine Hayhoe (an atmospheric scientist and evangelical Christian) describes the consequences of an extreme European heat wave in 2003:

> Over some parts of France, temperatures at the peak of the heat wave were almost 20°F warmer than average. Even worse, the killer heat persisted day after day, with no relief at night. With a population ill prepared to cope with extreme heat, the death toll soon became overwhelming. Bodies piled up in the morgues, the flower markets, the butcher shops. Every refrigerated surface available was used to store the bodies of nearly twenty thousand dead in France and over fifty thousand more in surrounding nations.[25]

This is not a scene from a dystopic horror movie; this happened in the heart of Western civilization with twenty-first-century resources.

On a global level, care for the neighbor involves care for those who will be most affected by a changing climate, populations that, historically speaking, contributed the least to the rise of rampant CO_2 emissions. Cynthia Moe-Lobeda, a Lutheran professor of Christian ethics, writes compellingly about the effects of climate change on the global poor as "a harrowing theological problem":

> The primal act of God—creation—is not merely to create a magnificent world. This God creates a magnificently life-furthering world. The scandalous point is this. We are undoing that very *tov* [Hebrew for "good," or "life-furthering"], undoing Earth's life-generating capacity. We—or rather, some of us—have become the "uncreators." Indeed, one young and dangerous species has become a threat to life on Earth. The credible scientific community is of one accord about this basic reality, and hundreds of its widely respected voices have been articulating it for over two decades. Less widely accepted, however, is a corollary point of soul-searing moral import. The horrific consequences of climate change are not suffered equally by Earth's people. Nor are the world's people equally responsible. Those least responsible for the Earth's crisis are suffering and dying first and foremost from it.[26]

She concludes with a powerful call to moral action:

> Our moment in time is breathtaking. It is pivotal. The generations alive today will determine whether life continues in ways recognizably human on this beautiful and broken planetary home called Earth. May Christians bring the gifts of our faith traditions to the great moral-spiritual challenge of the twenty-first-century—forging ways of living that Earth can sustain and that build justice among people. Doing so will mean holding raw anguish and joy in one breath. It will

25. Hayhoe and Farley, *Climate for Change*, 85.
26. Moe-Lobeda, "Climate Injustice," 532–33.

mean seeing good and evil tangled up together, with no person or system being either all good or all bad. And it will mean savoring the sensuous delights of life in this good garden Earth, while letting holy rage serve the call to love.[27]

This is the pressing moral issue of our time, and yet an issue with no real precedent, and therefore a tangled issue still without clear solutions and a clear theology.

A CHRISTIAN VOCABULARY FOR CLIMATE CONVERSATIONS

To return to my exchange with Bishop Gutierrez: the bishop continued to say that he didn't think that the Episcopal Church discusses climate change very much because there wasn't much teaching on the subject, and therefore we don't know how to talk about the problem—we lack a vocabulary. Interestingly, this dovetails with how we talk about climate change in our culture more broadly. American attitudes towards climate change have been shifting. An extensive study from the Yale Program on Climate Change Communication in April 2019 revealed that only 8 percent of the population does not believe that climate change is occurring. Growing majorities now recognize the threat of a changing climate: 62 percent of Americans are "somewhat worried" about climate change, and 23 percent are "very worried" about it. Additionally, 38 percent say that they have been personally affected by climate change, 48 percent believe their families and/ or communities will be harmed, 69 percent believe that future generations will be harmed, and 71 percent believe that plant and animal species will be harmed. People are worried about climate effects in their local area like extreme heat (69 percent), droughts (64 percent), flooding (60 percent), and/or water shortages (59 percent). But while 64 percent of Americans say the issue of global warming is "extremely," "very," or "somewhat" important to them, an equal percentage say they "never" or "rarely" discuss the issue with family and friends. Only about half (51 percent) of people say they hear about climate change in the media at least once per month, and only about a quarter (23 percent) say that they hear people talking about the subject at least once per month.[28] Taken together, these statistics demon-

27. Moe-Lobeda, "Climate Injustice," 540.

28. The statistics in this paragraph are all taken from the Executive Summary for the Yale Program on Climate Change Communications study "Climate Change in the

strate that as a society we feel that climate change is a serious concern, but we aren't discussing it.

Thus we have a rapidly intensifying problem with profound moral and environmental consequences about which we are largely silent. We either don't want to talk about the problem (maybe it will just go away, or go somewhere else?) or we don't know how to talk about the problem (where to even begin?). In terms of talking and thinking about climate change—a massive, multidimensional problem, one that encompasses the social, the political, the economic, the technological, the personal, and the spiritual— it is once again helpful to make distinctions. There is one conversation to be had about how to address the problem itself—how to turn off the flow of CO_2, how to turn off the faucet of the metaphorical bathtub, and how to mop up the consequences that have already occurred and that will continue to occur even as we ratchet down our carbon emissions. (Sadly, there is already irremediable loss, and there will be more loss, probably much more, before we restore balance.) Then there is another conversation to be had about how to think about the ethics of the problem.

Here, then, are some broad suggestions for how we can frame a conversation about the ethics of climate change:

- First and foremost, think of the issue of climate change as part of our religious and spiritual life. In the Yale study cited above, only 9 percent of Americans consider climate change as a religious issue. For people of faith, folding climate change into religious life brings the concern within a familiar moral and biblical context, and the readings on the moral compass thus become much clearer.

- Embrace the s-word. Liberal Protestant Christian traditions are sometimes embarrassed to use "sin," finding it old-fashioned and judgmental; conservative Protestant Christian traditions tend to use the term for specific personal actions, like sexual transgressions. But labelling our wanton collective burning of carbon, and our gratuitous individual carbon emissions, as sin prompts us to confront the harm of our actions, to face the fact that we have been crossing the road to avoid the sight of the battered person or planet lying by the wayside.

- Differentiate between Sin and sin, between the original sin into which we were born (that is, a consumer culture and a capitalist system that historically has been indifferent to its cost of burning carbon) and the

American Mind: April 2019."

ecological harms caused by our personal actions. We have an obligation to address both carbon Sin and sin, but making the distinction can help us to think more clearly about the different types of actions we need to take, and how we take them on a social and personal level.

- Differentiate between livelihood and luxury emissions. On a micro level, this enables me to think of making cookies with my kids as a livelihood emission, but carelessly pre-heating my oven for an hour while I answer some e-mails is a wasteful luxury emission. On a larger level, this makes me think harder about my patterns and mode of travel, or even the patterns of travel within my profession. It is unrealistic, I think, to ask people at this moment to get to zero carbon emissions, but there are choices that can be made in all facets of our life.

- Consider sins of commission and sins of omission, what we have done and what we have left undone. Which of our actions are contributing to climate change? What actions are we not taking to reduce CO_2 emissions?

- Use your words: justice, compassion, mercy, righteousness, stewardship, responsibility, etc. Different Christian traditions place different emphases on how we are to follow through with the central Christian tenet to love God and love neighbor. Whatever the mode of thinking and speaking in your tradition, it aligns with conversations about climate change and the ethical imperative to love one another. Discussions about the climate, and what we can do to improve it, do not have to be foreign to a community's discourse, but can be folded into familiar ideas.

- Live the gospel message: be doers, not just hearers. It is not enough to listen to the message about climate change; it requires action. And it requires evangelism; "[f]ewer than half of Americans perceive a social norm in which their friends and family expect them to take action on global warming."[29] Spread the word, shift the cultural norm, share the good news that with individual and collective actions we can mitigate the worst effects of a changing climate.[30]

29. Yale Program on Climate Change Communications, "Climate Change in the American Mind."

30. The national climate report concludes with a graph and a chart that illustrate strikingly different scenarios (across a range of sectors including roads, West Nile virus, wildfires, etc.) for a future with no climate action and one in which steps have been taken for mitigation and adaption. The human, environmental, and economic costs without

Let's turn now to these actions.

BECOMING DOERS, NOT MERELY HEARERS

The first action is to face the problem honestly and fully. We need to acknowledge and accept the reality and the scope of the climate crisis. The cautionary words of Pope Francis might sting a bit:

> [W]e can note the rise of a false or superficial ecology which bolsters complacency and a cheerful recklessness. As often occurs in periods of deep crisis which require bold decisions, we are tempted to think what is happening is not entirely clear. Superficially, apart from a few obvious signs of pollution and deterioration, [we might think that] things do not look that serious, and the planet could continue as it is for some time. Such evasiveness serves as a license to carrying on with our present lifestyles and models of production and consumption. This is the way human beings contrive to feed their self-destructive vices: trying not to see them, trying not to acknowledge them, delaying the important decisions and pretending that nothing will happen.[31]

Self-congratulatory pats on the back about virtuous recycling are woefully insufficient. Tackling climate change will need to happen on three fronts: (1) at the level of technology; (2) at the level of policy; (3) at the level of personal action.

Moving towards a lower level of carbon emissions (or even, as a recent report from the United Nations Intergovernmental Panel on Climate Change advocated, a *negative* level of carbon emissions[32]) is going to require a range of technological developments. Renewable energy (solar, wind on land and sea, geothermal) will need to scale up; battery storage is going to need to be improved; carbon sequestration probably needs to be part of the picture; power grids need to be redesigned; infrastructure for electric vehicles needs to be rapidly expanded. Again, I am not a scientist, and I have very little technical knowledge of how any of these actually work. But I have great hope, because of the research I see happening at my own university and around the world. There are myriad ways in which

---

mitigation are huge. U.S. Global Change Research Program, *The Climate Report*, 170.

31. Pope Francis, *Laudato Si'*, 41.

32. De Coninck et al., "Strengthening and Implementing the Global Response," 342–46.

brilliant, dedicated people are applying themselves to innovation and solving problems.[33] The younger generations are fully aware of the challenges and consequences of climate change, and they are laboring to develop solutions. From what I can see, previous generations approached problems with a vision of large-scale unified systems, such as building the interstate highway system. By contrast, in our own refracted world (we no longer even watch the same television shows through the same platforms, for instance, or get news of current events from the same sources), there is not going to be one single systemic shift, but diverse and manifold changes in how we generate, distribute, and use energy.

At the policy level, there is now widespread agreement among economists that the single most effective lever we can pull to reduce our collective CO_2 emissions is to put a price on carbon and let market forces do their work.[34] Carbon pricing in effect works with, not against, some of the neoliberal economics that Northcott describes as being the very source of the problem. As long as carbon is cheap (and still heavily subsidized[35]), and as long as the true cost of carbon's pollution and environmental impact is not factored into its price (such as the costs of health damage resulting from the burning of fossil fuels, or the billions of dollars we spend on cleanup and repair after increasingly destructive natural disasters[36]), people will continue to use it. Putting a fair, realistic price on carbon levels the playing field of

33. To take one example of innovation that blew my mind, the state of Delaware (working with scientists at the University of Delaware) has passed legislation that allows electric cars to connect to the grid so that the cars *feed into* the power supply; rather than thinking of cars as energy consumers, we can think of them as individual power generators; see Roberts, "Driving Clean Energy Forward."

34. See, for instance, Akerlof et al., "Economists' Statement on Carbon Dividends."

35. Hescox and Douglas lay out some striking statistics. Citing an article from *Forbes*, they note that from 1994 to 2009 there were about $477 billion dollars in subsidies for fossil fuels, versus $6 billion for renewables, not even including the tax breaks for fossil fuels (tax laws that date back to 1916). "Moreover, the highly respected International Energy Agency (IEA) reported that fossil fuels are securing $550 billion a year in subsidies worldwide and holding back investment in cleaner forms of energy. Oil, coal, and gas received more than four times the $120 billion paid out in incentives for renewables, including wind, solar and biofuels"; Hescox and Douglas, *Caring for Creation*, 116–17, citing the IEA's "World Energy Outlook 2014."

36. For a concise account of how climate change affects health care costs, see the National Resources Defense Council's policy statement "Health and Climate Change," or the World Health Organization's fact sheet, "Climate Change and Health." For a snapshot of the costs of climate-related disasters in 2018, see Banis, "10 Worst Climate-Driven Disasters of 2018."

the market so that technological innovations have a chance to come into their own. The most politically viable and socially equitable form of carbon pricing is that of Carbon Fee and Dividend (CFD), a revenue-neutral carbon price that returns the monies collected from the fee directly to the population.[37] This model of CFD has been promoted by the nonpartisan Citizens' Climate Lobby as well as a number of conservative and progressive organizations.[38] One way to be doers and not just hearers is to urge Congress to support and enact such a policy. Recognizing that the spiritual call to action on climate change also necessitates coordinated collective policy, the national Episcopal Church, as well as other faith organizations, have endorsed a CFD model.[39]

In terms of the personal, reducing our collective carbon usage will be the inverse of littering the street. If everyone throws just a little bit of trash on the street—a wrapper, a can, a newspaper, what have you—soon the streets are polluted. If people don't throw their trash on the street, the streets don't become polluted. Even if only half the people stop throwing their trash on the street, the pollution becomes half as bad. It is a question, in many ways, of changing collective behavior. This can happen. In one of the most infamous scenes in the AMC television show *Mad Men*, the picture-perfect Draper family goes on a country picnic on a beautiful

37. An internet search for "The Basics of Carbon Fee and Dividend, Citizens' Climate Lobby" leads to a clear diagram of the model (https://citizensclimatelobby.org/basics-carbon-fee-dividend/). Essentially, the model has three components: 1) Place a steadily rising fee on fossil fuels (assessed at point of entry into the economy—well, mine head, or port). 2) Give 100% of the fees minus administrative costs back to households each month (this covers the cost of rising prices for most families, and incentivizes a move away from a carbon-intensive lifestyle). 3) Use a border adjustment to stop business relocation (so that companies don't just move to places that have cheaper fossil fuels). In 2019, a bill that would implement the CFD model (with some modifications, like ways to help farmers), the Energy Innovation and Carbon Dividend Act (H.R. 763), was introduced in the U.S. House of Representatives, https://www.congress.gov/bill/116th-congress/house-bill/763 and https://energyinnovationact.org/ (accessed December 10, 2019).

38. For conservative groups, see the Climate Leadership Council and Students for Carbon Dividends, largely comprised of college Republicans; see Ted Halsted's TED talk on YouTube, "A Climate Solution Where All Sides Can Win," https://www.ted.com/talks/ted_halstead_a_climate_solution_where_all_sides_can_win?language=en

39. The Episcopal Church passed a resolution at the 79th General Convention in 2018 in support of CFD; see the policy statement at https://episcopalchurch.org/posts/ogr/eppn-creation-care-series-carbon-tax. As of the time I am writing this, CFD has also been endorsed by the Presbyterian Church (USA) and the Committee on Domestic Justice and Human Development of the U.S. Conference of Catholic Bishops; evangelical authors Hescox and Douglas also support CFD, see *Caring for Creation*, 134.

summer day in the early 1960s. As they pack up to leave, the lovely Betty Draper calmly shakes out the picnic blanket over the grass, leaving all their trash while the family drives away in a sparkling clean car. The camera lingers on the bucolic but now litter-strewn scene for a full half minute, letting the environmental sin sink in. Betty clearly has no sense of behaving inappropriately, but the torrent of posts on a YouTube clip of the scene indicate how social mores have shifted. What had once been an acceptable action has now been redefined as a violation of cultural norms. (And along with this redefinition, the excuse that one person throwing their litter on the street justifies my own littering no longer holds merit.) Or, to give another, first-person example: when I first moved into my Philadelphia neighborhood, most people didn't clean up after their dogs, and walking down the sidewalk was like navigating a minefield. But the code shifted, and now (most) people automatically pick up their dog's messes. This will seem like a trivial example in the face of potentially catastrophic climate change. But the point is that the path to lowering our collective CO_2 emissions requires both macro transformations (the large-scale reductions that can be prompted by pricing carbon, technological innovations in renewable energy development, and, ultimately, a re-tooled economy that lives within the means of the planet's natural resources) and micro transformations (the hundreds of small decisions we make in the course of a day), and that changing behaviors and social norms can happen, and does happen.

A GIFT FOR THE FUTURE

Actions follow attitudes. To shift our behaviors, then, we need to change how we think about our actions. Because carbon usage has been part of the human-created habitat we were born into, it is often hard to place carbon emissions into our pre-existing moral framework. Carbon emissions have been largely invisible and off our moral radar. So we need to first teach ourselves to be more cognizant of our direct and indirect carbon usage, and then consider how we can think about reducing our carbon emissions— how we can frame the problem/ethics/remediation of climate change to ourselves and others.

One popular mode of thought is to position climate change in a discourse of dieting, one of the modern ways our culture depicts virtue and self-control. For instance, a book that gives tips on reducing your own carbon footprint is titled *Low Carbon Diet: A 30 Day Program to Lose 5000*

Pounds.[40] Another links the personal and the planet, with *Go Green Get Lean: Trim Your Waistline with the Ultimate Low-Carbon Footprint Diet;*[41] a quick internet search of "low carbon diet" reveals that this linkage of personal food consumption and reducing global CO_2 emissions is a popular one.[42] "Just like going on a diet, cutting your carbon emissions is not always easy," writes the founder of a sustainable community.[43] Like many people, I have been absorbing the message, and have in fact been adjusting my family's food choices—at the store I now reach for the locally grown produce, and on our table fish and poultry have replaced beef, by far the most carbon-intensive meat. (I'm now trying to cut back on cheese, but boy, do I love good cheese.) But as a long-term motivator, the dieting analogy is not going to work for me, and I would guess it won't work for most people. Diets come and go in fads, and they rarely have permanent effects. If the dieting analogy inspires some people to reduce their carbon usage, that's great, but deprivation rarely motivates.

I am more likely to be spurred into action by thinking about the ethical dimensions of climate change as a positive matter of my faith and my spiritual life. Thinking about CO_2 emissions in the context of sin might seem dire (or dour), but having recognized that our invisible emissions have consequences for the planet and its people, we can then re-imagine our acts that reduce CO_2 emissions as acts of grace, as acts that serve God and others. CO_2 reduction is then not conceived in the restrictive language of dieting and self-control, but in terms of care-taking and gift-giving. If putting CO_2 into the environment—through that which we have done by using carbon, and that which we have left undone through denial, willful ignorance, or obliviousness of the costs of carbon usage—is sin, reducing our carbon use is an act of redemption.

If acts of carbon reduction are acts of gift-giving, for whom is the gift? There are many ways to consider this. We give to the local and the global neighbor, as expressed above by Moe-Lobeda. And we clearly give to the earth, thereby nurturing our own spiritual relationship with the world around us. Caring for the neighbor and the earth are obviously powerful

40. Gershon, *Low Carbon Diet.* The Amazon website for this book cites a review from *Plenty* that describes the book as "a step-by-step program, à la Weight Watchers, designed to reduce a person's carbon footprint."

41. Geagan, *Go Green Get Lean.*

42. See, for instance, the Bon Appétit Management Company's page, "Tackling Climate Change Through Our Food Choices."

43. Sirna, "How You Can Go on a Carbon Diet."

concerns, and they matter to me deeply. In this book, though, I want to focus on a particular kind of caring for the neighbor: caring for the people of the future. In our culture today, centered largely on the now, it is perhaps easier to think of taking care of people from other places rather than people from other times, especially the people who will come after us. But if we accept that reducing CO_2 emissions is an ethical necessity, and if we conceptualize our efforts at these reductions as an act of radical generosity that will help others, it is important that we develop a clearer way to speak of those in the future. Casting our minds to the future, however, is a counterintuitive impulse given the ways that our culture tends to position our historical moment in terms that emphasize the present. This privileging of the present results from the temporality required by consumerism. To re-quote from Northcott, "Consumer culture promotes an obsession with the present, and with satisfaction derived from present experiences, and a correlative loss of a sense of the significance of both history and posterity."[44]

The way that we conceptualize time is both a cause and a consequence of the modern world. Like many others, I have grave concerns about how our culture thinks about time. Ours is becoming a culture with an aversion, antipathy, or amnesia about history. Learning about history has been de-centered in the educational curriculum. With the decline of teaching and thinking about the past comes an attendant decline in thinking about oneself as a historical agent, someone who has inherited a legacy of past thought and action, and someone whose thoughts and actions will in turn impact the future. On many fronts, we seem to have become a culture of presentists. We can see this in political terms ("Gratitude to the past and obligation to the future are replaced by a nearly universal pursuit of immediate gratification"[45]), in economic conditions, and in environmental choices. Recently the *New York Times* ran a couple of stories about Kenya within weeks of each other. One article outlined the devastation to an agrarian community now faced with the frequent and enduring droughts that result from climate change.[46] The other reported on the construction of a Chinese-backed coal-burning power plant near that very region, noting the difficulties that activists had in convincing their peers of the damage

---

44. Northcott, *Moral* Climate, 93.
45. Deneen, *Why Liberalism Failed*, 39.
46. Sengupta, "Hotter, Drier, Hungrier."

such a plant inflicts upon the environment: "Compared to the compensation money on the table, their warnings about the future seem vague."[47]

That vagueness is my concern here. In particular, we lack a way of talking about the rights of the people of the future, and our ethical obligation to them. The West developed a robust set of actions for honoring those in the past, ranging from medieval prayer guilds for the dead to monuments to biographies. We often value historical preservation, and the law has enshrined the rights of the dead to specify their bequests of property. But beyond a general expression of concern for future generations, we do not really have an ethical vocabulary for talking about those who will live after us. And this lack of discourse is part of a lack of engagement with future human beings that is in some ways integral to Christianity. While the Bible and centuries of Christian ethics have contributed to a robust sense of obligation to the "Other" or the "neighbor," we do not have a comparable discourse for people of the future. This absence can be seen as the legacy of the "Christian embarrassment": early Christians understood that Christ's return to earth would be imminent, and there was a growing cultural and theological unease as the years (then decades, then centuries) passed and the Second Coming didn't occur. Given the expectation of imminent return, and perhaps the stigma of weak faith to think that it might not happen for some time, the crucial, foundational years of Christian theological formation did not contend with human futurity. As I will discuss in the next chapter, this has now rendered the "Christian embarrassment" an embarrassment in the literal sense of the word: as a blockage. Our missing theology of the future hinders our ways of talking about the ethics of climate change. If we look at the photo of the lean women that accompanies the *New York Times* article about the consequences of extended droughts in Kenya, we can recognize a Christian call to care for and respond to the global neighbor—and indeed, the article emphasizes the international contributions of food aid that have come to the region. The article, though, is not just outlining a problem of the present, but also implicitly pointing to the suffering of those in the future.

What should we even call "the people of the future"? One biblically-based term (that I discuss in chapter 3) might be "generations": if we look for future people in the Bible, we are most likely to find them in the promises to generations to come. This term, though, is perhaps too constricted to blood lines and family to carry the wide ethical sense that is needed for a

47. Sengupta, "Why Build Kenya's First Coal Plant?"

global crisis. Although the urge to provide for the future of one's own prog-
eny is more compelling than a generalized "people of the future," we need to
define a temporal relationship that is the analogue to the spatially-rendered
concept of "neighbor." I would propose that a biblical term we might adapt is
"inheritor." Jesus preached, "Blessed are the meek, for they shall inherit the
earth" (Matt 5:5). With "inheritor" I hope to capture a number of valences:
the sense of familial line that is a more intimate circle of care (our children
are, quite literally, our heirs); the wider notion of inheritance promised by
Jesus in the Sermon on the Mount; the importance of the "meek," since
those who will be most profoundly affected by climate change will be the
global poor; and, of course, the phrase "inherit the earth," if not originally
carrying the ecological weight that "earth" does today, aptly sums up the
planetary consequences of our actions. Within the Bible, we read of the
earth as our inheritance from God—Solomon says to God of his people,
"you teach them the good way in which they should walk; and grant rain on
your land, which you have given to your people as an inheritance" (1 Kgs
8:36)— and the people of the future will inherit the earth that we, in turn,
give to them.

This book, then, seeks to open conversations on how we might better think about our inheritors. The following chapters do not consistently
engage with climate change *per se*. My intention is to spur thoughts about
futurity that will motivate the moral actions needed to curb climate change
and the devastation it brings. I have called the reduction of personal CO_2
emissions an act of radical generosity, because in many ways we are not
acting simply to ensure a better life for ourselves or for our immediate children. Taking steps to mitigate climate change is an act of ethical courage
and an expression of care for the as-of-yet unnamed and unborn multi-
tudes who will take our place.

In a larger context, this book participates in a cultural conversation
about reframing how we think and talk about climate change. The psychologist and economist Per Espen Stoknes notes, "If we are able to reframe the
climate issue . . . in our society's discourse, there is less fear and guilt attached to it—more a sense of 'collective efficacy' or the idea that we can do
something together as a society. Now, this deep reframing of the issue takes
time—it's different than simply having a slogan or a new news headline. But
reframing impacts how people feel about and perceive the issue."[48] Stoknes

---

48. Quoted in Suttie, "How to Overcome 'Apocalypse Fatigue.'" See my discussion of
Stoknes's book on pp. 11–12 of this book's introduction.

touches on religious reframing—the shift from thinking in terms of dominion to stewardship of creation—but his solutions are primarily secular. He emphasizes the importance of societal influence in directing our actions. He references a famous study in which four groups of a thousand households each were given different motivators for reducing their use of power: the first group was asked to conserve power because it was right for the planet; the second group was asked to conserve for future generations; the third group was told how much they would save on their utility bill if they conserved power; and the fourth group was told how their energy use compared to that of their neighbors.[49] Were people motivated to reduce their energy consumption out of concern for the environment or future generations? No; the prime motivator was finding out how much energy the neighbors used. Society is a powerful shaping force. I am advocating that we fold care for the inheritors into our spiritual practices as a way of reframing our attitudes and actions towards climate change, but this is not mutually exclusive with positive societal influence. Indeed, as members of the church—as a people in communion—we can find a personal and a social focus on future generations to be mutually reinforcing. Thinking intentionally about the inheritors as we consider the environmental impact of our own actions can be part of a larger theological orientation and collective Christian care for the future.

Biblically speaking, the destruction of nature is the fault of a people, not a person: again, "the earth will be desolate because of its inhabitants, for the fruit of their doings" (Mic 7:13). We are all the inhabitants who leave the earth to our inheritors. Let the fruit of our doings, the products of our actions, be not environmental desolation, but a flourishing future earth. Let us remember that this is still possible through our choices, through being in right relationship with God's creation and people. We read of environmental reversal in the Psalms, of destruction returning to natural abundance:

> [God] turns rivers into a desert,
> springs of water into thirsty ground,
> a fruitful land into a salty waste,
> because of the wickedness of its inhabitants.
> He turns a desert into pools of water,
> A parched land into springs of water.

49. Suttie, "How to Overcome 'Apocalypse Fatigue.'" The study being referenced is Vladas Griskevicius, Robert B. Cialdini, and Noah J. Goldstein, "Social Norms: An Underestimated and Underemployed Lever for Managing Climate Change," IJSC 3 (2008) 5–13.

And there he lets the hungry live,
And they establish a town to live in;
They sow fields, and plant vineyards,
And get a fruitful yield.

. . . .

[H]e raises up the needy out of distress,
and makes their families like flocks.
The upright see it and are glad;
And all wickedness stops its mouth.
Let those who are wise give heed to these things,
And consider the steadfast love of the LORD.

(Ps 107:33–37; 41–43)

The threatened—and, increasingly, realized—environmental consequences of climate change are a product of our "wickedness," in the psalmist's terms, of our collective ignorance, luxury, and consumer desires. But "considering the steadfast love of the Lord," trusting in humanity's better angels, and working—with God's help—towards a solution, the fruitful vineyards can still lie before us.

2

Embarrassment

The Second Coming and the Stunted Christian Ethics of Futurity

Grant your people grace to love what you command and desire what you promise; that, among the swift and varied changes of the world, our hearts may surely there be fixed where true joys are to be found.

—Episcopal Book of Common Prayer[1]

THE ABSENT PRESENCE OF THE FUTURE

Christianity has long had a future problem. By this I don't mean that we have a problem in the future—although climate change clearly will be a problem in the future—but that we are limited in the way that we can talk about the human future and the moral rights of future human beings. We grapple to place our inheritors in a familiar discourse, a familiar epistemology or way of knowing the world. Nicholas Stern positions future people in a discourse of rights and social justice, folding "discrimination by date of birth" into a civil rights context that advocates against social prejudice and

1. Collect for the fifth Sunday in Lent (Year A).

political oppression: "Discounting future welfare or lives means weighting the welfare or lives of future people lower than lives now, irrespective of consumption and income levels, purely because their lives lie in the future . . . This is discrimination by date of birth, and is unacceptable when viewed alongside notions of rights and justice."[2] I came across this quotation in a chapter written by the evangelical author and meteorologist Paul Douglas, who himself places climate change within an anti-abortion or right-to-life discourse: "[Concern about climate change] isn't about polar bears. *This is about the health and welfare of our kids, and their kids, and their kids. . . .* Respect for the unborn must extend to future generations of the unborn. Climate change is a global pro-life issue."[3] And this passage appears in a section where Douglas is implicitly trying to move the conversation about the science of climate change out of a discourse of faith: "Do I *believe* in climate change? This isn't about belief or opinion. This is about acknowledging the data, taking time to understand the science, being mindful of what's happening worldwide, connecting the dots, and making decisions based on the best available evidence."[4]

These three rhetorical moves—contextualizing climate change in a discourse of social justice; contextualizing climate change in pro-life discourse; and shifting climate change from a discourse of faith/belief to one of factual data and scientific analysis—happen in rapid sequence within a single paragraph, leading to a wild moment of rhetorical and cognitive ping-pong. Taken together, they illustrate both how important it is to match the ethical problem of climate change to the right discourse, and how awkwardly we are equipped to talk about people of the future. While we can slide our inheritors into other conversations about rights (here, in the context of abortion or discrimination), the people of the future largely do not have their own discourse. This matters hugely, since a fundamental aspect of human communication is the necessity of genre and particularized realms of discourse. Genre and discursive realms give our conversations shape and familiarity, and thereby meaning.

On the whole, we are trained to think inside of the box, or rather, inside of different boxes. Categories are what allow us to create meaning.

2. Stephanie Kirchgaessner, "Moral Case to Tackle Climate Change Overwhelming, Says Lord Stern," *The Guardian*, September 10, 2015; cited in Hescox and Douglas, *Caring for Creation*, 91.

3. Hescox and Douglas, *Caring for Creation*, 91, original italics.

4. Hescox and Douglas, *Caring for Creation*, 90–91.

Consider movies: if I call a particular film a Western, or a horror movie, or a romantic comedy, or science fiction, or a documentary, or any other well-known genre, you can immediately imagine the elements (lighting, soundtrack, types of camera shots, dialogue) that typically mark that genre. Of course, some of the most interesting art is that which plays with the boundaries of its box, perhaps mixing in elements of another genre, but with rare artistic exceptions we need the box to help us mentally process what we see and read and hear. These boxes enable us to talk to each other meaningfully; if we are having a conversation about Westerns, everybody knows what we mean.

Recognizing cultural and intellectual boxes in all of their variety (types of cars; types of food; types of sports; types of clouds; etc.) is how we come to know the world. Raising children involves teaching them normative cultural categories. When my daughter was three years old, she had a stack of playing cards with animal pictures on them. One day she was sitting on the floor with her cards, studiously sorting them into two piles. She would take a card from the deck, study it intently, and decide which pile it belonged to. Intrigued, I quietly watched for a while trying to figure out the code, the basis on which she was separating the cards. I couldn't make any sense of her system, and finally asked her what the different piles were for. She looked at me incredulously, surprised that it wasn't obvious, and simply pointed to the piles and said, "Big noses, small noses." I don't remember how I learned how to sort animals, but at some point I was naturally conceptualizing them in terms of mammals, birds, insects, and reptiles. In high school we refined this further, sorting into kingdom, phylum, class, order, family, and genus. The idiosyncratic epistemological categories of children are charming and sometimes startling, and it is a bit sad to see them grow out of them, but to be a functional adult in our society you need to be working within (or, intentionally against) the normative systems of classification. If you go to the zoo or a restaurant or biology class assuming that animals are differentiated by nose size, you will have difficulties.

Today of course it is fashionable to think outside of the box, but this is not a call to abolish boxes themselves—as individuals and as a civilization we could not function without operating categories. Rather, to think outside the box is to recognize the boxes, to acknowledge how they channel and often constrain our thoughts and behaviors, and to notice what or who doesn't fit comfortably in any one category. (We can think here of the famous case of the platypus, an egg-laying mammal.) What can be really difficult to identify, though, are the boxes we don't have, the categories that do not exist in our culture. What categories do we lack? What are we missing? Often we can sense an invisible or ghost category, something that we don't yet see, when we struggle for words or labels. And sometimes the invention of a label can spontaneously make visible a box for which we only had a ghostly impression. Think of terms like "suburban sprawl," "yuppie," "selfie," "climate refugee." Before there was the label, there was no specific category or concept, just an awful lot of houses gobbling up cornfields, or a group of work-obsessed young people discovering French cheese and Italian coffee, or people taking pictures of themselves, or people trying to get away from devastating drought.

In Western Christianity (the kind I know—I won't speak for others), there has been a missing conceptual box for the people of the future. It is this category that I am trying to call into being through labeling it with "the inheritors." Without the label, there is not a classification; without a classification, it is difficult to have conversations and make decisions about attitudes and actions. See, again, Paul Douglas's moment of discursive grappling—he is reaching for a steady way to talk about our ethical response to *"our kids, and their kids, and their kids,"* his italics calling attention to the importance of the problem, but also inadvertently revealing that we don't have a clear way to talk about the kids of the kids of our kids.

Why don't Western Christians have a clear way of talking about the kids of the kids of our kids? Or the kids of the kids of other people's kids? Why didn't a Christian discourse and a vocabulary for future human beings develop over the last two thousand years? Part of the reason lies in the sacred text that anchors the religion. Within the Bible, Jesus promises to return, perhaps soon after his departure. The expectation of Christ's imminent return is fundamental to the origins of Christianity, resulting in a Christian temporal framework in which the start of the religion is also always already the end. Simply put, from Christianity's inception there was not a developed sense of future human beings because it was widely assumed that there would not be many future human beings. Ethically, spiritually, and communally speaking, there was no need to consider human futurity because the end of human history was near. In the beginning was the end, at least as some early interpretations of Christian Scriptures are concerned.

In this chapter, then, we will consider the tricky issue of Christ's Second Coming—the Parousia—in terms of how it has laid out Christian discourse about the future, and how it preempted a working category for human beings of the future, *our kids, and their kids, and their kids.* (As I will discuss in the next chapter, the Hebrew Scriptures engage extensively with future human beings through the category of "generations," a category that is largely cut off in the New Testament. In the Psalms, for instance, we read of "the steadfast love of the LORD . . . and his righteousness to children's children, to those who keep his covenant" [Ps 103:17–18]; the Christian Scriptures don't look that far into the generational future.) First, we will look at the biblical assertions of the Parousia, which are notoriously complicated and even appear contradictory in terms of a timeline. Even within the New Testament, the proclamations of Christ's imminent return to earth are undercut by vague or more temporally distant expectations and Jesus's own words that he does not know when the return will take place. Second, we will consider how both the non-event (thus far) of Christ's return and the periodic claims of an imminent Parousia have become a source of embarrassment for many Christians (especially mainline Protestants and some evangelicals), resulting in a tendency to shy away from discussions of the future altogether. Finally, we will turn to contemporary theology as a way to embrace both the promise of the Parousia and human futurity, considering how the very embarrassment of the Parousia—its deeply unsettling sense of temporal indeterminacy—can lead us towards a new way of thinking about our inheritors.

The threat that climate change presents to the people of the future compels us to practice neighborly care for them, but it is difficult to enter into a caring, ethical relationship with a group of people who have no category—especially, in the case of our inheritors, a group of people who have been excluded from the realms of Christian theology. The hope of this chapter is to gain a deeper understanding of what has happened to the people of the future in the long course of Christian theology, and to find a way to bring the inheritors back into the realm.

THE TIME OF THE PAROUSIA

There are many biblical moments which address the Parousia, or the Second Coming of Christ: this is a central element of the Christian Scriptures. We can start with Matthew. In this gospel, Jesus prepares his disciples for their life of evangelism, giving them very specific instructions. They are not to go to the towns of the Samaritans or the Gentiles, but those of "the lost sheep of the house of Israel" (Matt 10:6); they are to cure the sick, raise the dead, cleanse the lepers, cast out demons (Matt 10:8); they are not to take payment (Matt 10: 8–10). It will not be pleasant; Jesus sends them "like sheep into the midst of wolves" (Matt 10:16), and "Brother will betray brother to death, and a father his child, and children will rise against parents and have them put to death; and you will be hated by all because of my name" (Matt 10:21–22). The life is to be one of perpetual movement ("When they persecute you in one town, flee to the next" [Matt 10:23]), but, comfortingly, this will not last forever: "for truly I tell you, you will not have gone through all the towns of Israel before the Son of Man comes" (Matt 10:23). The disciples' peripatetic motion thus also becomes a form of time-keeping, since Christ will return before they have been to all of the towns.

Matthew's Jesus details what will happen at the Parousia (a darkened sun and moon, stars falling from heaven, the Son of Man arriving on clouds from heaven, angels sounding trumpets, the elect gathered from the four winds—offering a pastiche of Hebrew prophecy[5]) and when it will happen: during the lifetime of "this generation" (Matt 24:34), the demonstrative Greek pronoun of *hautē* [αὕτη, meaning "this"] leaving little exegetical wiggle room for temporal uncertainty.

This proclamation is the culminating point of a heated diatribe that Jesus launches against the Pharisees, a passage that is often mined for moral imperatives but that also offers a rich and complex temporal setting. Indeed, Jesus almost seems to bait the Pharisees with their historical understanding and their place in history. He asks them who is the father of the Messiah, and when they answer that the Messiah is the son of David (Matt 22:42), Jesus stumps the Pharisees by asking, "If David thus calls him Lord, how can he be his son?" (Matt 22:45). The conundrum here appears to be one of hierarchy (how can the son be higher than the father?), but also one of time: David foretells of a Messiah (Ps 110) who is already "Lord." How can a son, who is inherently in the future, be the Lord, who is in the present and always has been in the past? Past, present, and future incomprehensibly converge. (We might picture the Pharisees dumbstruck, faced not with the textual problem they expected but a temporal one they did not: "No one was able to give him an answer, nor from that day did anyone dare to ask him any more questions" [Matt 22:46].) In a similar mode of temporal "gotcha," Jesus quotes the Pharisees' own self-justification—"'If we had lived in the days of our ancestors, we would not have taken part with them in

---

5. Jesus cites from Isaiah 13:10, Ezekiel 32:7, Ezekiel 33:4, and Joel 2:10, 2:31, and 3:15.

shedding the blood of the prophets'"—and uses it to condemn them: "Thus you testify against yourselves that you are the descendants of those who murdered the prophets" (Matt 23:30–31).

The Pharisees thus try to create historical distance between themselves and the past, but Jesus collapses that distance, connecting past deeds against the prophets with those who live in the present. Jesus condemns the Pharisees, declaring that "upon you may come all the righteous blood shed on earth, from the blood of righteous Abel to the blood of Zechariah son of Barachiah, whom you murdered between the sanctuary and the altar" (Matt 23:35). The present Pharisees are now the "you" who murdered the son of Adam and the last martyr in Hebrew Scripture, their mingled blood (and that of all of the righteous who were murdered in between) collapsing time. The temporal distancing that allows for chronological progression is obliterated, as the discrete identities of historical figures (the Pharisees and their fathers; Abel and Zechariah) merge in an instant. After this expansive and omnipresent temporality, the next verse is almost shocking in its historical specificity: "Truly I tell you, all this will come upon this generation" (Matt 23:36).

These warnings appear to make the disciples nervous, for later on, "When [Jesus] was sitting on the Mount of Olives, the disciples came to him privately, saying, 'Tell us, when will this be, and what will be the sign of your coming and of the end of the age?'" (Matt 24:3).[6] Jesus proceeds to warn them of false signs, to tell them of the true signs, and to claim (somewhat disconcertingly) that he himself does not know the exact hour of his return (Matt 24:36), but urges his followers to be alert and prepared for the end (Matt. 24:42, 44). Of crucial import here, though, is the fact that the Parousia is supposed to happen within the lifetime of that generation. It is also important that this is not a subtle or an ambiguous assertion, or a passing remark. Earlier, Jesus had declared, "Truly I tell you, there are some standing here who will not taste death before they see the Son of Man coming in his kingdom" (Matt 16:28). There is an unavoidable immediacy to this remark: some of the people standing right here, right now, will be alive to experience the Parousia.

---

6. Dr. Crystal Hall points out to me that the "end of the age" here needs to be read in the context of Roman imperialism, as a counter message against the hegemonic imperial message and also as a phrase that situates Matthew in a post-70 CE context (i.e., after the destruction of the Temple in Jerusalem). Her reminders about the post-70 CE context also inform my readings of Paul's epistles, below.

Borrowing from the apocalyptic genre, Jesus sketches a terrifying picture of this event: there will be war, famines, and earthquakes; the disciples will be tortured; there will be lawlessness, which will cause love to go cold (Matt 24:7–12). It will be a moment of terror, and spontaneous flight; the people will flee to the mountains; there will not be time to get things from the house, or to run back from the fields to grab a coat; it might happen in the cold of winter, or when people are unprepared on a Sabbath (Matt 24:16–20). And again, we find a mixing of temporalities. Jesus situates the Parousia in the long duration of cosmic history: "For at that time there will be great suffering, such as has not been from the beginning of the world until now, no, and never will be" (Matt 24:21). And yet, the Parousia will also be an event of temporal immediacy, without duration: "For as the lightning comes from the east and flashes as far as the west, so will be the coming of the Son of Man" (Matt 24:27). And in the midst of this general cataclysm, Jesus singles out only one particular group of people for special pity, the new mothers: "Woe to those who are pregnant and to those who are nursing infants in those days!" (Matt 24:19). It is a striking and exceptional moment. There is profound empathy here: in the general scene of panic and haste that Jesus has created, there is almost a pause to imagine the pragmatic difficulties of fleeing while pregnant or with a small nursing infant. There is an emotional projection here: in an image that has been made all too powerfully real in the globe's recent refugee crises, we can picture mothers desperately trying to protect their children and the unborn. Yet there is also a temporal focus here: Jesus zooms in, as it were, on the threshold of human futurity, on those who have just been born, or who are just about to be born. But it is the end. Woe to them.

This emphasis on the nearness of the Parousia is not limited to Matthew's Gospel. In Mark, Jesus says, "Truly I tell you, there are some standing here who will not taste death until they see that the kingdom of God has come with power," and "Truly I tell you, this generation will not pass away until all these things have taken place" (Mark 9:1, 13:30). And in Luke, Jesus says, "But truly I tell you, there are some standing here who will not taste death before they see the kingdom of God"; and "Truly I tell you, this generation will not pass away until all things have taken place" (Luke 9:27, 21:32). Written in the wake of the Jewish war and the destruction of the temple in 70 CE, the Gospels can express a sense of calamity and the end times. But the earlier Pauline epistles (ca. 55 CE) are even more urgent in expressing the imminence of the Parousia: "the appointed time has grown short," writes Paul, "For the present form of this world is passing away" (1 Cor 7:29, 30). And in the letter to the Romans: "you know what time it is, how it is now the moment for you to wake from sleep. For salvation is nearer to us now than when we became believers; the night is far gone, the day is near" (Rom 13:11–12). The moment is now.

We can imagine how such words would have filled early Christians with the weight of dread, and the thrill of hope. We can imagine (although perhaps with difficulty) how such an expectation might have shaped their daily lived experience. (Will Jesus return during the night? Before I have used up the olive oil I bought at the market? Before the springtime comes again?) We can imagine this expectation of the Parousia as a defining element of early Christianity. But: the pregnant women continued to give birth, the nursing infants continued to grow up, men became old and died. And more were born, and more died. Time marched on. "This generation" which was to have witnessed the Parousia faded; more took its place, and those faded, too. And more, and more. And still Christ hasn't returned.

THE PAROUSIA AND CHRISTIAN EMBARRASSMENT

Let's return to Matthew 24:34, "Truly I tell you, this generation will not pass away until all these things have taken place." Actually, let's return to an interesting footnote for this verse. R. T. France, in the New Bible Commentary, offers this: "The NIV [New International Version] mg. offers 'race' as an alternative to *generation*. This suggestion is prompted more by embarrassment on the part of those who think v 30 refers to the Parousia rather than by any natural sense of the word *genea*!"[7] To switch in "race" for "generation" is to take a big textual and interpretive liberty (as France points out—with exclamation point for emphasis—the Greek word *genea* [γενεά] does not mean "race"). Fascinatingly, France attributes this translation decision to "embarrassment." What is there to be embarrassed about? For some commentators, the issue is that Jesus said he would return very shortly after his death, and he did not—or at least not in the dramatic way that he so colorfully described. This fact is perceived as an embarrassment that touches the very heart of Christianity.

7. New Bible Commentary (University and Colleges Christian Fellowship, 1994, 1953; electronic text hypertexted and prepared by Oak Tree Software, Inc., version 2.0), in Accordance Bible Software, version 12.3.4

Indeed, the word "embarrassment" is frequently found in the context of discussions of the Parousia. We find this in Arthur Lewis Moore's *The Parousia in the New Testament* (1966), which fully explores the implications of the biblical references to the Second Coming, including possible internal textual evidence that the non-event of the Parousia was providing fodder for contemporaries who ridiculed the nascent religion.[8] The books of the New Testament were written at different points in time, of course, and so we can end up with multiple temporal frames even within a single book—an author is recording Jesus's life and words even as the passage of time has allowed for historical perspective on that story. Thus authors can write about the promise of the Parousia even as they also defend or qualify earlier accounts of the Parousia that were already attracting mockery.[9] Moore notes that "Lk. 9.27 is also understood by a number of recent scholars as evidence that Mk. 9.1 was causing acute embarrassment in the early church."[10] Second Peter 3, in particular, seems addressed to those who pointed to Christ's non-return to disparage the Christian beliefs:

> First of all you must understand this, that in the last days scoffers will come, scoffing and indulging their own lusts and saying, "Where is the promise of his coming? For ever since our ancestors died, all things continue as they were from the beginning of creation!" They deliberately ignore this fact, that by the word of God heavens existed long ago and an earth was formed out of water and by means of water, through which the world of that time was deluged with water and perished. But by the same word the present heavens and earth have been reserved for fire, being kept until the day of judgment and destruction of the godless. But do not ignore this one fact, beloved, that with the Lord one day is like a thousand

8. Moore, *Parousia in the New Testament*, 151–57.

9. For an example of early ridiculing of the Christian notion of the Parousia, see the comments of the second-century Greek philosopher Celsus: "They [the Christians] postulate, for example, that their messiah will return as a conqueror on the clouds, and that he will rain fire upon the earth in his battle with the princes of the air, and that the whole world, with the exception of believing Christians, will be consumed in fire. . . . It is equally silly of these Christians to suppose that when their god applies the fire (like a common cook!) all the rest of mankind will be thoroughly scorched, and that they alone will escape unscorched—not just those alive at the time, mind you, but (they say) those long since dead will rise up from the earth possessing the same bodies as they did before. I ask you: Is this not the hope of worms? For what sort of human soul is it that has any use for a rotted corpse of a body? . . . it is nothing less than nauseating and impossible"; Celsus, *On the True Doctrine*, 77, 86.

10. Moore, *Parousia*, 130.

years, and a thousand years are like one day. The Lord is not slow
about his promise, as some think of slowness, but is patient with
you, not wanting any to perish, but all to come to repentance. (2
Pet 3:3–9)

Of this passage Moore observes, "A number of scholars further maintain
that the number of ideas brought together here reflects the writer's embarrassment at the situation and the views of the mockers (showing what a
great problem the community was facing)."[11] One such scholar was Ernst
Käsemann, who contended that "[t]he whole community is embarrassed
and disturbed by the fact of the delay of the Parousia, a fact naturally used
by the adversaries to bolster up their arguments."[12]

We even find the famed Christian author C. S. Lewis confessing his
own embarrassment at biblical descriptions of the Parousia. Lewis recognizes the centrality of the Parousia to Christianity, citing references to the
Second Coming in the Apostles' Creed, the book of Acts, and the Gospel of
Matthew: "If this is not an integral part of the faith once given to the saints,
I do not know what is."[13] Yet he recognizes the "modern embarrassment"
over the Parousia, tackling this problem in both theoretical and practical
terms.[14] He ventriloquizes an atheist argument about the Second Coming,
and responds to this argument in an unexpected way:

> "Say what you like," we shall be told, "the apocalyptic beliefs of the
> first Christians have been proved to be false. It is clear from the
> New Testament that they all expected the Second Coming in their
> own lifetime. And, worse still, they had a reason, and one which
> you will find very embarrassing. Their Master had told them so.
> He shared, and indeed created, their delusion. He said in so many
> words, 'this generation shall not pass till all these things be done.'
> And he was wrong. He clearly knew no more about the end of the
> world than anyone else."
> It is certainly the most embarrassing verse in the Bible.[15]

Lewis's comment comes as a bit of a jolt, although he then rhetorically pairs
this embarrassment over Christ's presumed mistake with Christ's own

---

11. Moore, *Parousia*, 155. Moore himself concludes that within the New Testament
the authors viewed the Parousia as near, but without delimiting a time frame, 148.

12. Käsemann, *Essays*, 170.

13. Lewis, "World's Last Night," 93.

14. Lewis, "World's Last Night," 94.

15. Lewis, "World's Last Night," 97–98.

ignorance about the time of the Parousia in order to emphasize Jesus's full humanity. Less theologically, Lewis depicts a scene of actual embarrassment in the context of the failed apocalyptic predictions of "poor" William Miller in 1843. Of the cultural phenomenon that surrounded Miller's proclamations of the date of the Second Coming, Lewis wryly comments, "Thousands waited for the Lord at midnight on March 21st, and went home to a late breakfast on the 22nd followed by the jeers of a drunkard."[16]

It might not come as a surprise that the fiery apocalyptic proclamations of Matthew 24:34 rub against the tweedy English sensibilities of a C. S. Lewis. What might be more surprising is the recent turn in certain evangelical circles against an emphasis on the judgment of the Second Coming. To be sure, the anticipation of an imminent Parousia is still part of American evangelical culture in a way that does not find a counterpart in Catholicism or mainline Protestant denominations. For instance, the doctrinal statement of Liberty University, founded by the Rev. Jerry Falwell and a center of evangelical education, includes this: "We affirm that the return of Christ for all believers is imminent. It will be followed by seven years of great tribulation, and then the coming of Christ to establish His earthly kingdom for a thousand years. The unsaved will then be raised and judged according to their works and separated forever from God in hell. The saved, having been raised, will live forever in heaven in fellowship with God."[17]

But elsewhere in the evangelical community, people who were raised believing that theirs was the generation that would experience the Second Coming are now reconsidering the apocalyptic emphasis. We can look to a blog post called "Embarrassed by the Parousia? Evangelicals and the Return of Christ" (posted on the site Patheos) by Roger E. Olson, Professor of Theology at Truett Theological Seminary at Baylor University.[18] Olson notes the mainstream evangelical avoidance of the Second Coming, even though this was a hallmark of the faith for earlier generations. Olson reflects, "When I was growing up in the 'thick' of the American evangelical movement, one of its hallmarks was confident expectation of and looking forward to the imminent return of Jesus Christ. Our hymn books, recorded and broadcast music and preaching, sermons, prayers and devotional literature were filled with the themes of eschatology—the end of the world as

16. Lewis, "World's Last Night," 107.

17. "Doctrinal Statement," Liberty University website, http://www.liberty.edu/aboutliberty/index.cfm?PID=6907 (accessed November 5, 2018).

18. Olson, "Embarrassed by the Parousia?"

we know it by the 'second coming of Christ.'" But now, "Aside from the occasional outbreaks of 'eschatology fever' among fundamentalists evidenced by popular books forecasting a year when Jesus will return, American evangelicals as a whole have all but dropped eschatology as a subject of songs, sermons and conversation. Why?" Olson proceeds to outline a number of reasons: "Christian complacency due to prosperity"; an over-emphasis on eschatology during the 1960s and the 1970s, a time of threats to Israel and of nuclear war; the end of the Iron Curtain in the late 1980s. He notes the cultural influences that shaped the religion of his early life: "Evangelicals near my own age well remember *The Late, Great Planet Earth* and 'Like a Thief in the Night' and 'I Wish We'd All Been Ready.' That book (and the genre it inspired), that movie (and the genre it inspired) and that song (and the fear it inspired) so saturated evangelical consciousness that, when the Parousia (return of Christ) didn't happen, we turned inward with our spirituality and focused our outward attention on politics." But this shift from a focus on the Parousia was not just prompted by political changes and shifting cultural phenomena—actual embarrassment over some of the fundamentalist emphasis on ideas like the "rapture" and the Left Behind book series also played a role in the shift. Olson notes that he is "dismayed by those fundamentalist Christians who focus too much attention on the 'rapture,' the 'Great Tribulation,' and the 'Antichrist.' They give biblical, Christian eschatology a bad name. And they are why most mainstream evangelicals have simply given up on the Parousia and eschatology." In response to a reader's comment on the blog post, Olson sums up his point: "most of our churches (non-fundamentalist evangelical) are simply embarrassed by the whole area of eschatology."

Within the comment section, we find many who write to share their sense of embarrassment. Dave W. agrees with Olson, writing, "Yes I am embarrassed by the Parousia by which I mean almost all the talk, exact sequence of events, exact time . . . about end times that I hear . . . BUT I am not embarrassed by Christ's second coming and the rule by Christ." Olson responds, "it seems you are not embarrassed by the Parousia but by what I called 'eschatology fever.'" To which Dave W. replies, "I know but when the topic of the second coming comes up I strongly tend to simply shut up and walk away. Typically it comes up with my wife's relatives who tend to be Pentecostals." And apparently Dave W. and his in-laws are not the only ones to experience friction. John Wylie comments, "one possible reason for the lack of preaching on the Parousia may be that there is such a difference of belief between the pastor and the congregation. For instance, I believe that what we are looking for is the 2nd coming of Christ, but several people in the congregation are looking for the Rapture." Olson remarks that the pastor needs to remind his congregation that the Rapture is "nowhere mentioned in the Bible." Randy Starkey then weighs in, noting, "I always cringe when I see the bare statement made that the rapture is nowhere mentioned in the Bible. It all depends on what one means by the rapture. Definition becomes important. If you mean Hal Lindsey's scenario, I agree. But many evangelicals use the term rapture related to passages from 1 Thess. 4, Mt. 24, and even 1 Cor 15." This prompts Olson to clarify, "I assumed people would know that I meant the pre-tribulational 'secret rapture' which is pretty much how 'rapture' is used now." As the comment section veers into a discussion of the rapture, it seems to get away from a frustrated Olson ("Again . . . as I say to many commenters here . . . stick to what I said. I did not endorse any of that. But the Parousia is a biblical theme that cannot rightly be ignored by people who consider the Bible their authority for faith and life. It doesn't have to include what you are talking about"). The exchange also produces what may be one of the world's greatest typos: "Paul seems to have expected Christ's rerun in his life time" (David).

We might not be able to watch reruns of Jesus, but we can see videos of the song that Olson mentions, "I Wish We'd All Been Ready" (1969) by Larry Norman.[19] With his signature long blonde tresses that endured into middle age, Norman sings of an apocalyptic vision that is part Parousia, part rapture. In a quiet, high-pitched voice and accompanied by gentle guitar chords, Norman depicts a world populated by the sinners who have been left behind after the rapture that accompanies the Son's return. The disappearances of the godly are sudden: two men are walking up a hill when one vanishes; a wife awakes to find herself alone after her husband is taken from the bed. For those left behind—for those who weren't ready for the Parousia—there is profound loss, regret, and lamentation. While the soothing tune and the recurring sentiment of the title lyric suggest communal sympathy, the piece is really one of warning, chastisement, and judgment. Indeed, the clash of content (with Vietnam-era verses about war and dying children) and the simple melody—punctuated with bridges of "la-la-la-la-la"—renders the song almost absurdist, but apparently unwittingly so. In terms of the musical quality, the color of the lyrics and, yes, the long blonde

19. Larry Norman, "I Wish We'd All Been Ready," YouTube, https://www.youtube.com/watch?v=Ep4Aj-kJkJM.

hair, it seems fair to say that the song, at one point an evangelical anthem of the Second Coming, can indeed look embarrassing.

APOCALYPSE HOW?

What are the consequences of this embarrassment of the Parousia, an embarrassment that, as we have seen, threads its way from New Testament authors to canonical Christian authors like C. S. Lewis to debates within the evangelical community? What are the consequences for how we think of a concept of history that extends across past, present, and future? And what are the consequences of this type of historical thinking for cultural conversations about climate change?

In a rather blunt way, the recent religious emphasis on the Second Coming (and/or the rapture, even if thoughtful evangelicals are careful to separate this from the biblical Parousia) has rendered long-term planning for the future moot. Or as Headless Unicorn Guy, one of the commentators on Olson's blog post, puts it, "In my case, after being burned BAD by The Gospel According to Hal Lindsay back in the Seventies (followed by 10 years of Rapture Scare flashbacks), it's much better for my mental health if I stay away from the Parousia. After that tunnel vision on 'It's All Gonna Burn', Creation Care and having a future is a lot more my speed. Because when The World Ends Tomorrow and It's All Gonna Burn, don't expect anyone to dare great things, make long-range plans, or appreciate the here & now."

The evangelical emphasis on the Second Coming has also led people to interpret severe weather events caused by climate change as signs of an imminent Parousia. When the West Virginia town of White Sulphur Springs experienced catastrophic flooding in 2016, many considered this part of a larger pattern of environmental change that heralds the return of Christ.[20] One inhabitant who lost her home, Kathy Glover, was asked what actions the town might take after the flood, and her response echoes the refrain of Larry Norman's song: "There's nothing really I can encourage or discourage, other than to encourage people to be ready for the Return."[21] An article on the flood notes that "From Matthew 24 to the Book of

---

20. Subramanian, "Seeing God's Hand." The tag line of this article reads, "An evangelical mountain town lost eight people to flooding from an extreme rain storm. Many residents see the Biblical prophecy of the apocalypse, and welcome it."

21. Subramanian, "Seeing God's Hand."

Revelation, [the townspeople] were primed for apocalypse," and that for residents, "The greater the collection of disasters, the closer the long-awaited return of Jesus Christ."[22] And indeed, there are striking parallels between the environmental destruction of the end times foretold in the Bible and the environmental destruction of climate change as forecast in scientific reports (predictions that have increasingly become realities even during the short time in which I have been writing this book, as the effects of global warming are accelerating faster than almost anyone expected[23]). One commentator summarizes the parallels thus:

> According to the Bible, as the end of times approaches, the waters will turn 'bitter,' ocean-dwelling creatures will die and 'on the earth, nations will be in anguish and perplexity at the roaring and tossing of the sea.' The sun, the moon and the stars will be obscured and then the sun will heat up and burn mankind. It is not a stretch to interpret these passages as a presage of actual environmental problems: water pollution and air pollution that obscures the atmosphere . . . , acidification of the oceans and the resultant destruction of coral reefs, global warming, rising sea levels.[24]

Thus the "Rapture Index" at RaptureReady.com tracks leading environmental indicators like famine, drought, plagues, wild weather, and floods in the hopeful expectation of an imminent Second Coming; the increasing acceleration of climate change is not a source of alarm, but of excitement and promise.[25]

If the people of White Sulphur Springs (the very name of which conjures a mélange of the good, the demonic, and the hopeful) look with confident optimism at climate change as a sign of the apocalypse promised in Revelation, in other circles the idea of a dire "climate apocalypse" runs rampant. On the one hand, some conservatives have derisively viewed ecologists as "a reincarnation of the fanatics of the Apocalypse," since "conservative evangelical communities in the United States have long been suspicious of ecologists, whom they suspect to be Communist agents in disguise or converts of some wrongheaded eastern religion."[26] (Or as one

22. Subramanian, "Seeing God's Hand."

23. Linden, "How Scientists Got Climate Change So Wrong."

24. Mouhot, "In Pursuit of the Apocalypse," 6.

25. Scherer, "Christian-right Views Are Swaying Politicians."

26. Mouhot, "In Pursuit of the Apocalypse," 6–7.

climate denier puts it, "green is the new red."[27]) Thus even as some climate skeptics can attribute environmental changes to the imminent Parousia, the notion of anthropogenic climate change as "apocalyptic" becomes a source of conservative ridicule. On the other hand, those on a loosely-defined political left do indeed earnestly turn to "apocalypse" as a label to express the severity of the climate problem and to heighten dire warnings. This right/left tug-of-war over the term "apocalypse" is reflected in the media through article, book, and blog titles. From the right: "Is it time to cancel the climate apocalypse?"; "Not the Climate Apocalypse"; *Roosters of the Apocalypse: How the Junk Science of Global Warming is Bankrupting the Western World.* From the left: "We Can't Stop the Climate Change Apocalypse Through Individual Action Alone"; "Fuck your apocalypse: Between denial and despair, a better climate change story"; "Canadians end up the big losers in the apocalyptic climate change debate." And the term also peppers general analysis of this phenomenon: "Apocalypse Forever? Post-Political Populism and the Spectre of Climate Change"; "Apocalypse How? What Novels Screw Up About Climate Change"; "How Big Business is Hedging Against the Apocalypse"; and "Alien apocalypse: Can any civilization make it through climate change?"[28]

Historically, apocalyptic belief has been a part of Western Christian culture since the Gospel of Matthew (and even earlier, since the Christian apocalyptic vision draws on the deep well of end-of-days imagery in the Hebrew Scriptures and the genre of apocalypse from the inter-testamental period). While the topic of the Second Coming and apocalypse comes and goes in terms of Christian trends, it is an integral part of the religion's history for the past two millennia.[29] This belief in an imminent Parousia

27. Steve Milloy, *Green Hell: How Environmentalists Plan to Control Your Life and What You Can Do About It* (Washington, DC: Regnery, 2009), 234; cited in Collomb, "The Ideology of Climate Denial in the United States," 6; see 5–6 for discussion of climate change as a presumed socialist conspiracy.

28. Editorial Board (*Richmond Times-Dispatch*), "Is It Time to Cancel the Climate Apocalypse?"; Editorial Board (*Wall Street Journal*), "Not the Climate Apocalypse"; Isaac, *Roosters of the Apocalypse*; Rousseau, "We Can't Stop the Climate Change Apocalypse"; Thanki, "Fuck Your Apocalypse"; Harper, "Canadians end up the big losers"; Swyngedouw, "Apocalypse Forever?"; Williams, "Apocalypse How?"; Barron, "How Big Business is Hedging Against the Apocalypse"; University of Rochester Newscenter, "Alien apocalypse," which summarizes an article in a scientific journal, "The Anthropocene Generalized: Evolution of Exo-Civilizations and Their Planetary Feedback," *Astrobiology* 18.5 (May 1, 2018), https://www.liebertpub.com/doi/10.1089/ast.2017.1671.

29. For some representative titles, see Court, *Approaching the Apocalypse*; Palmer,

has proven itself to be "protean and resilient," a pervasive—if, for some, embarrassing—part of Western history that we tend to ignore:

> Christopher Columbus makes more sense in the modern age as an intrepid businessman (which he was not) trying to make a profit, than as the dupe of medieval apocalyptic theory (which he was) trying to be a prophet. Similarly, modern historians of science pass discreetly over Isaac Newton's efforts to date the end of the world and discover alchemical formulas, preferring to present him as the Titan of modern science rather than the millennial magus of his own culture's idiom.[30]

It is beyond the scope of this chapter to provide a historical survey of apocalyptic beliefs, although the interested reader could turn to the accessible *Apocalypses: Prophecies, Cults, and Millennial Beliefs through the Ages.*[31]

With over 2,000 years of very weighty baggage, then, the popular connotations of the apocalypse become an impediment to developing the future-oriented ethics and spirituality needed to address climate change. It is in this sense that the Parousia can present an embarrassment according to a meaning the word held from the seventeenth through early twentieth centuries: according to the *Oxford English Dictionary*, an "embarrassment" is "Something (material or immaterial) which is a hindrance or encumbrance; an impediment, obstruction, or obstacle; a difficulty, a problem" (def. 1.a.). The Parousia is so central to the New Testament that it is disingenuous for Christians to ignore it. At the same time, it is so clear that our failure to address climate change imperils billions of future human beings—this is a profound ethical failure, a refusal to show the type of mercy and compassion that Jesus teaches. How then can we resolve this fundamental conundrum of keeping faith in an imminent Parousia while also maintaining an ethical commitment to care for the people of the future?

The Apocalypse in the Early Middle Ages; Whalen, *Dominion of God*; Backus, *Reformation Readings of the Apocalypse*; Bell, *Apocalypse How?*; Johnston, *Revelation Restored.*

30. Landes, "Fruitful Error," 89–90.

31. Weber, *Apocalypses*. For a more detailed book that explores the "historic framework for . . . apocalyptic anxieties" (12), see Friedrich, *The End of the World*. Friedrich—notably writing in 1982, before climate science had become mainstream—sees the fears of natural catastrophe (including the melting of the polar ice caps) that appear in "underground newspapers" as carrying forward the Cold War-era fears of nuclear annihilation (11).

THE NOW AND FUTURE KINGDOM

Contemporary theology can perhaps lead us out of this quandary. Although in some ways, the sense of embarrassment over the Parousia might seem to persist even here.[32] For instance, the German theologian Jürgen Moltmann writes, "For modern theology the early Christian expectation of the Parousia is an embarrassment which it thinks it can get rid of with the help of demythologization."[33] But others have found a way forward by boldly embracing the very notion of embarrassment. I turn to a section of David W. Congdon's book *The God Who Saves: A Dogmatic Sketch* (2016) entitled, "Salvation as Embarrassment: Eberhard Jüngel's Eccentric Eschatology." ("Eschatology," again, is the branch of Christian theology that addresses "last things," or the end of worldly human history.) Congdon engages with Jüngel's theology, and especially Jüngel's distinctive approach to the Parousia, as part of his larger engagement with the theology of salvation (i.e., soteriology). Congdon notes how Jüngel "reject[s] the notion that the non-occurrence of the Parousia undermines all future eschatology, not because the Parousia will actually arrive in some distant future but because a literal, chronological understanding of the Parousia misunderstands the function of imminent expectation for the life of faith."[34] In other words, thinking of the Second Coming as an event that happens along a sequential timeline misses the point. According to Jüngel, responses to biblical accounts of the Parousia have erred in one of two directions: the negative explanation that early Christian expectation was a mistake "misses the essence of imminent

---

32. A summary of theological approaches to the Parousia over the last century can be found in Hayes, *When the Son of Man Didn't Come*. In the introduction to the volume ("Was Jesus Wrong About the Eschaton?"), Hayes provides a historical survey of the vacillation concerning the Parousia—from Johannes Weiss and Albert Schweitzer reviving interest in the eschatological expectation of Jesus in the early twentieth century, to the Jesus seminar's rejection of Schweitzer's view in the 1980s and 1990s, to the current re-revival of eschatology (4–8). The authors agree with Schweitzer that Jesus prophesied the Parousia would occur shortly after his earthly ministry, but the non-event of the Parousia should not undercut Christian hope; rather, "the delay of the Parousia is entirely consonant with the way ancient prophecy works and the operations of the God that Christians worship" (18). But the book's blurb by Anthony C. Thiselton returns us to embarrassment: "This is an outstanding book. It skillfully combines not only biblical studies and Christian theology, but also liturgy and hermeneutics to show that the delay of the Parousia, far from being an embarrassment, enlivens the hope of the Church."

33. Jürgen Moltmann, *The Way of Jesus Christ*, 313; cited in Chester, *Mission and the Coming of God*, 13.

34. Congdon, *God Who Saves*, Loc. 2101.

expectation"; the positive explanation that the Parousia has been delayed "eliminates, through the positivity of its explanation, *the embarrassment that the problem of imminent expectation poses and must pose.*"[35]

Jüngel's approach to embarrassment essentially takes what might be considered a bug of the system and turns it into a feature. What is the "essence of imminent expectation"? Congdon answers:

> According to Jüngel, it is *embarrassment.* To be sure, there is embarrassment in the fact that an expected parousia does not occur, as many a fundamentalist cult could confirm. But Jüngel's point goes deeper than that. He seems to suggest that the expectation of Christ's imminent arrival is an intrinsically unsettling hope, one that fills us with mounting unease and discontent. The doctrine of imminent expectation is not a problem that demands a definitive solution, either through rejection or affirmation, but instead a belief whose purpose is to *problematize the believer.* Ironically, therefore, the nonoccurrence of the parousia actually *fulfills* the purpose of the expectant hope insofar as it compounds the embarrassment and thereby actualizes the essence of Christian eschatology.[36]

Congdon's (and Jüngel's) use of "embarrassment" here is not the edge of the word that veers into shame or avoidance. Rather, they are thinking of embarrassment as it leans into a sensation of being unsettled or disconcerted. This unsettling disallows pat chronological resolutions—the Parousia did not take place, or the Parousia will take place—and instead renders the Parousia as more of a productive state of spiritual and theological disorientation. Here again is Jüngel via Congdon: "In contrast to a negative or a positive way of explaining the problem of imminent expectation and the delay of the parousia, *it is dogmatically essential to keep alive the theological embarrassment that is given with this problem.*"[37] Congdon continues, again working with Jüngel, "The removal of this embarrassment . . . is just as problematic as a 'chronological fixing' of the parousia. The certainty that comes

---

35. The quotation here is from Jüngel, thesis 4.31, in Congdon, *God Who Saves,* loc. 2114. Original italics.

36. Congdon, *God Who Saves,* Loc. 2130.

37. Congdon, *God Who Saves,* Loc. 2141. Congdon is citing Jüngel, "Der Geist," 313 (thesis 4.322); the emphasis is Congdon's. For ease of reading I have deleted from this quotation the theologians associated with the negative or positive explanation; Jüngel cites Rudolf Bultmann as an example of the negative mode of the thought, and Karl Barth as an example of the positive.

with either denying or affirming the coming advent leads to the comfortable domestication of the kerygma [i.e., proclamation of the Gospel]. To prevent this, the embarrassment 'must remain . . . a lively embarrassment,' so that we never experience ourselves as settled and secure."[38]

Having grounded his thinking in Jüngel's theology, Congdon then offers his own contribution to ways of thinking about the Parousia, incorporating embarrassment into a way of being Christian. His thoughts are worth quoting at length:

> If the essence of imminent expectation is its embarrassment—if the parousia is fulfilled precisely in its delay, in the unresolvable dissonance with the present moment—then the hope in a 'definite future' and the expectation that the glorified Christ is 'near at hand' cannot mean a hope in some literal, public occurrence in the distant future. But that does not mean a definite future is mere wishful thinking or that Christ is not, in fact, near at hand. It means that we must rethink the nature of the promised future and of Christ's nearness—not to lessen in any way the embarrassment of the parousia's nonoccurrence, but to understand how this nonoccurrence can itself be the fulfillment of Christ's promise. If Christ is near at hand in his very disruption of our propensity toward easy solutions and self-security, then his parousia occurs in the event of the word of justification that disturbs our illusions of peace and safety and thereby places us outside ourselves (*extra nos*).[39]

For Congdon, then, eschatology is about a radical de-centering of our egos, about getting beyond ourselves or getting over ourselves, as the saying goes. "Jesus's message of the coming kingdom and Paul's message of God's justifying righteousness are both different ways of proclaiming the message of a reality that resides outside of ourselves, a reality that, when we encounter it . . . , places us outside of ourselves as well. The eschatological word makes us eschatological creatures. The eschaton, the apocalypse of salvation, comes near to us as we become distanced from ourselves."[40]

Congdon's answer to "rethink[ing] the nature of the promised future and of Christ's nearness" is to decenter ourselves, to make ourselves excentric to salvation history. This destabilization is expressed in similar ways by Rowan Williams (the former Archbishop of Canterbury of the Anglican

38. Congdon, *God Who Saves*, Loc. 2141. The interior quotations again come from Jüngel, "Der Geist," 313 (thesis 4.322).

39. Congdon, *God Who Saves*, Loc. 2141–2153.

40. Congdon, *God Who Saves*, Loc. 2165.

Church), who writes of God as "the event that attacks and upsets my self-image," a breaking through that

> brings on a kind of vertigo; it may make me a stranger to myself, to everything I have ever taken for granted. I have to find a new way of knowing myself, identifying myself, uttering myself, talking of myself, imaging myself . . . . [I struggle] for a discipline that stops me taking myself for granted as the fixed center of a little universe, and allows me to find and lose and re-find myself in the interweaving patterns of a world I did not make and do not control.[41]

This breaking-in is Christ, who "assumed the right to declare, on God's behalf, that our history, our world, the drama of our egos, was at an end in the presence of God."[42] *This* is apocalyptic. And indeed, Congdon gives another name to his eccentric eschatology: "I have defined that which is 'outside ourselves'—what the reformers identified as Christ, grace, faith, and the word—in the terms of the *future*."[43]

Martin Luther also translated *parousia* as "Zukunft," choosing the German word for "future" rather than the word for "return"—"but future in the sense of something coming towards the present rather than something developing (becoming) out of the present."[44] This future, then, is outside of or extrinsic to our own temporality. It is something over which we have no control; it is outside of our time and space, our systems, and our selves. We move towards this future, even as it is coming towards us (as the German preposition "zu" [towards] in "Zukunft" indicates). In this eschatological motion we encounter future (and past) generations.

And here we face another embarrassment, another obstacle presented by the Parousia, or particular ways of understanding the Parousia: if the expectation of imminent return has cut off contemplation about future human beings and our obligations to them, the notion of an eccentric

---

41. Williams, "A Ray of Darkness," 101, 100.

42. Williams, "A Ray of Darkness," 103.

43. Here is the full quotation, which places this idea firmly in a reformed tradition: "An eccentric eschatology is simply the historical-hermeneutical reconstruction of the reformational particles: *solus Christus, sola gratia, sola fide, solo verbo*—by Christ, grace, faith, and the word alone. The particles amount to one simple claim: *the truth of our existence lies outside of ourselves.* By reconstructing soteriology in light of the eschatological context of early Christianity, I have defined that which is 'outside ourselves'—what the reformers identified as Christ, grace, faith, and the word—in the terms of the *future*." Congdon, *God Who Saves*, footnote 187, loc. 2153.

44. Chester, *Mission and the Coming of God*, 12.

eschatology, or a return that breaks open the self—as valuable as the concept is—threatens to replace a literalist fear of individual judgment with a self-focused soteriology (theology of salvation), even if the self is that which must be annihilated or overcome. How can we hold in our minds, or in prayer, an understanding of a divine breaking in (and breaking of) ourselves that keeps this self in relationship with community? How do we contemplate a joining of self with neighbor(s), and a concern for this community on both a spiritual and a material level—that is to say, how do we think of eschatological union with others, as well as a forward-looking care for future human bodies and selves? To make an obvious but crucial point: in the Bible, Christ frequently speaks of his coming in terms of the collective, in terms of the society of "kingdom" or "realm." We can look to two moments in Luke: "Once Jesus was asked by the Pharisees when the kingdom of God was coming, and he answered, 'The kingdom of God is not coming with things that can be observed; nor will they say, "Look, here it is!" or "There it is!" For, in fact, the kingdom of God is among you'" (Luke 17:20–21). And, "As they were listening to this, he went on to tell a parable, because he was near Jerusalem, and because they supposed that the kingdom of God was to appear immediately" (Luke 19:11).[45]

As these passages also illustrate, running through Christian Scripture is seemingly contradictory language about the arrival of this kingdom/realm: "the kingdom of God is [already] among you," yet the disciples err in their supposition "that the kingdom of God was to appear immediately." This paradox of the fulfilled and the unfulfilled— a "tension between 'already' and 'not yet'"—marked the life of the early church.[46] According to Moore, this tension was "[n]ot between certain End events which have been accomplished and certain others which have not yet been fulfilled, but between the End events fulfilled in a mystery already (fulfilled, that is, in the hidden ministry of Christ), and the manifestation of their fulfillment in openness which has not yet occurred."[47] For early Christians, the Parousia was not just a future event, but a way of understanding their place in a larger eschatological time: "It is in [an] understanding of past and present centred on Christ and mediated to us through the Spirit, that the early

45. The Greek word that is here translated as "kingdom," *basileia* [βασιλεία], indicates the spatial concept of land perhaps better captured in "realm," which carries less of the emphasis on political structure in "kingdom." Thanks to Dr. Crystal Hall for pointing this out to me.

46. Moore, *Parousia*, 170.

47. Moore, *Parousia*, 170.

church has found itself compelled to live in imminent expectations of the End."[48] This expectation, it could be said, lights a fire under the early communities, for "the opportunity for mission is temporary, and therefore that the missionary task of the church is urgent, forbidding idle wistfulness and lethargic sorrow."[49]

We find, then, not only a merging of the individual and the collective, and of the present and the future, but also of the *imminent* and the *immanent*—per the definitions of these two words in the *Oxford English Dictionary*, that which is "ready to befall or overtake us" or "close at hand," and that which "exists or operates within," is inherent, and "permanently pervades and sustains the universe (God)."[50] The *OED* notes that from the seventeenth through nineteenth centuries "imminent" was sometimes used for "immanent," and labels this a "confusion." But maybe it is not confusing at all: that which is temporally close at hand and that which is materially close at hand are, theologically, interconnected. If we consider the Parousia an existential event which moves us to "rethink the nature of the promised future and of Christ's nearness," and "if Christ is near at hand in his very disruption of our propensity toward easy solutions and self-security" (again, Congdon's language), then we are jolted out of ourselves to experience and account for what it means to be in God's kingdom. With the imminent as immanent, even familiar words become powerfully eschatological and expressions of the Parousia: "thy kingdom come, thy will be done, on earth as it is in heaven." Future kingdom/present planet/self/others collapse into one. The arrival of God's kingdom places us in the midst of "the interweaving patterns of a world [we] did not make and do not control," to return to Williams's words. Or, to use a more scientific notion, the arrival of God's realm puts us inside of an ecosystem, a structure of deep interconnection, one that weaves together the divine, the natural world, and the human— and humans across time. An ecosystem is comprehended as a totalizing relationship: all is interwoven and interdependent, and there is nothing outside of it. Our consumer culture peddles in the fantasy that individuals do not reside within a spiritual ecosystem, that we can opt out, that it is our privilege and prerogative and within our capacity to step outside of

48. Moore, *Parousia*, 171.

49. Moore, *Parousia*, 148.

50. I am adapting for quotation the *Oxford English Dictionary's* definitions of "imminent" (adj. 1) and "immanent" (adj. 1).

this ecology (or, to use trinitarian language, this economy[51]). But "Parousia" names the mystical ecosystem we inhabit, the integration of human and divine across time.[52]

Imagining the human beings of the future, then, need not entail a vision of waves of generations marching forward in linear time, away from us, as we watch their backs recede into a misty distant time. The Parousia brings those of the future into the eschatology of our present—that of the divine ecosystem, the divine ecology, the divine economy. The Latin term for the Greek *parousia* (παρουσια) is "adventus," an arrival. We can think of the inheritors not as distant, faceless Others, but as fellow citizens of the Kingdom of God, sharers of the realm who have arrived with Christ to be with us in our present; this can be their category. And we can take up the urgent sense of mission the Parousia inspired in the first Christians—building a community centered on the generous care of others. The task calls for deep thoughts as well as sensible shoes. We might give the last word to C. S. Lewis, who calls us to look to the Parousia but to avoid eschatology fever and simply get on with the work:

51. In chapter four I will discuss the conceptual homology of ecology and trinitarian economy.

52. My thinking here has been informed by the brilliant observations of Dr. John Hoffmeyer, shared in an e-mail on December 11, 2018: "Jesus' foundational proclamation and conviction that the basileia of God has drawn near could be twisted into a presentism that says, 'to hell with caring about the future.' But that would be to falsify the character of the basileia. To say that the basileia has drawn near is to say that a whole structuring and dynamic of life, of interconnected relationships, of ecology, of cosmos—specifically, a structuring and dynamic permeated by the governance of God—have already come near. The consequence is that to live at odds with that structure and dynamic, to live at odds with that governance, does not make sense. There is no time left to get ready for the time when it will make sense or be advisable to live in accordance with that structure and dynamic. That is, the reasons for putting off living in accordance with the governance of God have no firm foothold. In 2 Cor. 6, Paul, quoting older scripture, says: 'Now is the acceptable time; now is the day of salvation.' We could interpret this with two very different temporal valences. One is what you call 'presentism,' of the kind rampant in contemporary consumer capitalism: relative disregard for past and future; absorption in present gratifications of desires and/or present distractions against sorrows, pains, and anxieties. The other interpretation is to unmask the so-called presentism as an elaborate and deceptive mechanism for eluding living presently. We can not and/or will not live presently—which is always structured by pasts and futures—in the sense that we can not and/or will not accept being with the present reality, in its complexities, including temporal complexities, of joys and sorrows. So we seek to escape living presently by narcissistic self-gratification, distraction, and numbing—for which contemporary consumer capitalism not only provides endless purchasable opportunities, but for which contemporary consumer capitalism also fashions a spiritual ethos."

[P]erpetual trepidation about the Second Coming . . . will certainly not succeed. Fear is an emotion: and it is quite impossible—even physically impossible—to maintain any emotion for very long . . . . Crisis-feeling of any sort is essentially transitory. Feelings come and go, and when they come a good use can be made of them: they cannot be our regular spiritual diet . . . . Frantic administration of panaceas to the world is certainly discouraged by the reflection that 'this present' might be 'the world's last night'; sober work for the future, within the limits of ordinary morality and prudence, is not. For what comes is Judgment: happy are those whom it finds laboring in their vocations, whether they were merely going out to feed the pigs or laying good plans to deliver humanity a hundred years hence from some great evil.[53]

If this is somewhat flip theology, it is good life advice, especially for those engaged in climate work. Apocalyptic fear and fervor are unsustainable motivational forces, exhausting those who need to undertake "sober work for the future." But happy are those who labor in their vocations. Addressing the challenges of climate change requires daily acts of small charity towards our inheritors as we collectively lower our burning of fossil fuels, as well as courageous acts of leadership. In the spirit of the Parousia, we can understand these to be actions on earth as it is in heaven, forever and ever.

53. Lewis, "World's Last Night," 109, 111.

3

Across Time

An Intergenerational Environment and a Liturgical Response

[O God,] make us ever aware of the presence of this great company. Grant that we may find, in the reality of your nearness, the nearness of those countless other servants who are separated from us by years and distance.[54]

TIME SENSITIVITY

In a beautiful piece of photojournalism, Anne Barnard describes the great cedars of Lebanon:

> Walking among the cedars on a mountain slope in Lebanon feels like visiting the territory of primeval beings. Some of the oldest trees have been here for more than 1,000 years, spreading their uniquely horizontal branches like outstretched arms and sending their roots deep into the craggy limestone. They flourish on the moisture and cool temperatures that make this ecosystem unusual in the Middle East, with mountaintops that snare the clouds floating in from the Mediterranean Sea and gleam with winter snow.[55]

54. Rowthorn, *The Wideness of God's Mercy*, 102.
55. Barnard and Haner, "Climate Change is Killing the Cedars of Lebanon."

The article is accompanied by the powerful photography of Josh Haner, whose moving images take us into a misty forest of these mysterious trees, their craggy, lichen-covered bark and finger-like branch tips making them seem more like animals than vegetables, or at least more animated than vegetative. Over the wonder of the forest, we read the words, "The ancient cedars of Lebanon have outlived empires and survived modern wars. Clinging to shrinking patches of territory, these trees stand for Lebanon's resilience. Now, global warming could finish them off."

Figure 2. A cedar tree in the Maasser Cedar Forest, south of Beirut.
JOSH HANER/The New York Times/Redux.

There are a number of ways in which climate change is impacting the cedars, according to the article. The seeds of the cedars must freeze in order to germinate, and they like to have snow. Mild winters can thus inhibit the seeds' ability to germinate. And when seedlings do successfully sprout, they are increasingly being tricked by early spring days into coming up too soon, making them vulnerable to cold snaps that can freeze them to death later in the season. The cedars have been trying to inch their way up the mountain as they pursue the receding cooler temperatures, but the mountain only goes so high and the cedars only grow so fast. Eventually there could be no ground with the temperature required for germination, no place with the potentiality for the seeds to come into life. On top of all this, annual rainfall and the length of winter's snow cover have dropped off precipitously from what they were only a generation ago. The warmer temperatures have also spurred the sawfly, an insect that buries itself in the winter, to emerge earlier in the season, laying its eggs in time for the larvae to feast on young needles, thereby devastating forests. So while infant trees can be killed by a warm winter or spring, mature trees can die of thirst or a plague of devouring insects. What can we do to save the trees? Short of mitigating the rise in global temperatures caused by our burning of fossil fuels, there is little action that we can take. These trees, for all of their imposing physiognomy and longevity, have adapted to a particular climate and are unable to withstand delicate changes in the environment. Human ingenuity and technological advancement cannot save them.

I have long marveled at the beauty of the world, but I have only recently been learning just how precise are the steps and movements in nature's dance. Animals, insects, and plants have developed over deep time to relate to their environment in very specific and interrelated ways. I have been learning about this, in part, from my students. A few years ago, I was on my university campus during an unseasonably warm day in March. The air was lovely and temperate (in the words of Shakespeare) and in the low seventies (according to the thermometer). After a bitterly cold winter, the radiant sun felt glorious on the face; after months of wearing layers and boots and wooly mittens, it felt liberating to be outside in a thin blouse and light sweater. And after months of seeing students shuffle around campus, hunched against the wind in their dark coats, it was rejuvenating to see them in their many-colored t-shirts playing Frisbee or, like me, just sitting on a bench and opening themselves to the sun. The world seemed to be coming back to life, and life was good. I knew that the unseasonable temperatures were probably part of a shifting climate (this was March but it felt like May), yet on that day the weather was so wonderful it seemed as if climate change wouldn't be so bad after all. Later, back in my office, I met with a student who was passionate about birds and studying to become a research ornithologist. She was anxious, almost stricken. When I asked what was wrong, she told me that she was really worried about the weather—that with a sudden and early warm up, hibernating insects could start moving and flying about, and when the weather returned to freezing (as it almost certainly would, and in fact did) the insects would die. This sounded like a tragedy for the bugs, but not like a problem for me. Then my student said that when there is a massive insect die-off, the birds that rely on them for food also starve. I began to imagine my world without birds, and it was a more dismal place.

Maybe someone doesn't like bugs or birds, or even the great cedars of Lebanon, and couldn't care if they disappear. But most people like and expect to eat. I have been learning about the fragility of crops from my father, who has a small apple farm in western Michigan. As the weather has become more erratic, the region has seen some striking consequences of climate change; unprecedented torrential downpours, for instance, flooded a town cemetery to the point that coffins were pushed to the surface.[56] But less dramatically, when the steps in nature's dance are thrown off by the slightest of rhythms the result is crop failure. "Time sensitive" has become a common designation of priority in our workaday world, but we have lost sight of how sensitive time truly is in the world of nature. Some years ago, my father's apple orchard was in full bloom. There are few wonders that can rival a springtime orchard—gnarly branches erupt in a profusion of exquisitely delicate pink and white petals, pollinators orchestrate a gentle rustling hum, and the blossoms send forth a fragrance that saturates earth and skin. But if a freeze comes when the flower is at a particular stage of development, there won't be an apple—the blossoms, of course, are apples in their infancy. That year a warm spring had resulted in an early bloom, and then came a freakish late frost. The blossoms can withstand temperatures just below the freezing point and still survive to develop into an apple, depending on conditions like elevation and the duration of the freeze. But that night my father's orchard succumbed to the cold, and all was lost. ("Not even enough for pie," my father said of his crop that year. The neighbor's orchard, at just a slightly different altitude, did not sustain damage.) The loss of this crop for a small-time farmer is not the end of the world as we know it. But my father's apple trees and the great cedars of Lebanon illustrate why even small changes in temperatures and timing can have big consequences. And if we start to scale up the loss of an apple crop across a planet of rapidly shifting weather patterns (that is, a changing climate), we can see the potential for cataclysmic human suffering.[57] The words of

---

56. "Flood waters unearth caskets."

57. "'[F]ood production shocks' linked to climate change have been rising globally, putting food security at risk . . . [and] the frequency of crop production shocks driven by extreme weather had been increasing steadily. Food shocks threaten to destabilize the global food supply and drive up global hunger rates, which have started to tick up in recent years"; Gustin, "Investors Join Calls for a Food Revolution." (The scientific study cited by Gustin is Richard S. Cottrell et al., "Food Production Shocks Across Land and

Henry David Thoreau, in a chapter called "The History of the Apple-Tree," now seem ominously prescient: "It is remarkable how closely the history of the Apple-tree is connected with that of man."[58]

In reading about the dangers that climate change poses to the natural world, from the great cedars of Lebanon to the humbler apple to the smallest of insects, it is intriguing to note the biblical ring to many stories. In the article on Lebanon, we read that the most famous cedar patch is known as the "Cedars of God," a place where the resurrected Jesus is reputed to have revealed himself to his disciples.[59] And the phenomenon of global insect decline was described in a much-circulated piece called "The Insect Apocalypse Is Here," which notes the 2017 international headlines of an "insect Armageddon" after a study showed a 75 percent drop in the insect population of German nature reserves over a twenty seven year period (for midsummer populations, the drop was a shocking 82 percent).[60]

The author of "Insect Apocalypse," Brooke Jarvis, describes a phenomenon called "shifting baseline syndrome," in which subsequent generations are born into an environment that is increasingly degraded but they don't notice the damage, since one generation's degradation is the next generation's normal. Jarvis writes, "The world never feels fallen, because we grow accustomed to the fall." In *Paradise Lost*, John Milton's great seventeenth-century epic poem about the Fall of Adam and Eve, there is a poignant moment when Eve realizes that she must leave the flowers she knows and loves because she is being expulsed from Eden for breaking

---

Sea," *Nature Sustainability* 2 (2019) 130–37.) Other recent studies show the possibility of climate change leading to simultaneous "breadbasket" failures around the globe; Freedman, "Extreme Weather Patterns." (One of the scientific studies cited by Freedman is Kai Kornhuber et al., "Amplified Rossby Waves Enhance Risk of Concurrent Heatwaves in Major Breadbasket Regions," *Nature Climate Change*, December 9, 2019.) For a terrifying read, see McKibben, "This is How Human Extinction Could Play Out," an excerpt of his book *Falter: Has the Human Game Begun to Play Itself Out?* (New York: Henry Holt and Company, 2019). Less cataclysmic but still disturbing, see Severson, "From Apples to Popcorn"; the subtitle of this article reads, "In every region, farmers and scientists are trying to adapt an array of crops to warmer temperatures, invasive pests, erratic weather, and earlier growing seasons."

58. Thoreau, *Wild Apples*.

59. Barnard and Haner, "Climate Change is Killing the Cedars of Lebanon," quoting Antoine Jibrael Tawk.

60. Jarvis, Brooke. "The Insect Apocalypse is Here." Jarvis notes that the study was from "an obscure German entomological society"; I have not been able to locate the original study, but Jarvis notes that it was the sixth-most-discussed scientific article in 2017.

God's commandment.[61] She is heartbroken, and there is an overwhelming sense of sadness when she and Adam must leave the garden paradise for the last time. They truly experience loss, as the poem's title indicates. But beyond the narrative boundaries of the poem, we know that their sons Cain and Abel will be born into a harsher natural world. Perhaps Adam and Eve will tell their sons stories of their life in Eden, and how the world used to be. Perhaps Cain and Abel, like many children, will be vaguely interested in what their parents have to say, but will be more invested in their own realities. Perhaps children today will someday try to explain what snow was like to their own children. And yet, "The world never feels fallen, because we grow accustomed to the fall." If we become accustomed to the Fall—the loss of harmony between human and divine, and humans with one another—perhaps we also feel no postlapsarian guilt, no sense that we are those guilty of ecological sin. Although the Western pictorial tradition of the Fall depicts Eve with an apple in hand (versus the more generic "fruit" in the Bible), we resist any accountability for a crop that is too small to even produce an all-American apple pie.

The ethics in the biblical myth of Adam and Eve are straightforward: God gave a command not to eat the fruit of a certain tree, the command was violated, there were direct consequences. In a world that is ever-falling but where we don't experience or perceive the loss, the ethics become complex. We find ourselves in an intergenerational chain of environmental sin. No one person, and not even one generation, is responsible for environmental degradation. There is not one particular Fall of Man—nor one particular Fall of Nature (in *Paradise Lost*, the entire natural world falls slightly when Eve tastes the fruit[62]). The burning of coal in seventeenth-century London polluted the air; early "pilgrims" arrived on a wooded Cape Cod and managed to promptly and permanently destroy a delicate forest ecosystem; the Industrial Revolution put gashes in the earth; DDT made the rivers run toxic; industrial pollution caused the great Lake Erie to catch fire; the automobile filled Los Angeles (and now other cities around the world) with the choke of smog; we clog the oceans with plastics and continue to fill the atmosphere with excess carbon dioxide. And not only are the environmental ethics complicated by intergenerational agency and culpability, but much of the damage has been done unwittingly and sometimes even with good intentions: people wanted to heat their homes, or build homes, or

61. Milton, *Paradise Lost* in *Complete Poems*, Book 11, verses 268–279.
62. Milton, *Paradise Lost* in *Complete Poems*, Book 9, verses 782–84.

invent new machines to replace tedious human labor, or kill the mosquitos that cause malaria, or build great bridges, or enjoy a life of individual freedom, or live a comfortable life seemingly inextricably adapted to modern convenience. Unlike Eve's fateful reach for the fruit, we cannot point to one action done by one person; we bear a distributed burden.

If we are fully to come to terms with our role in anthropogenic climate change—that is, an environmental shift that has its genesis in *anthropos* (the human)—we need to place ourselves inside of a larger temporal landscape, a larger vision of human history. Recognizing how we are both indebted to and victims of past generations can help us to see how tremulously future generations are at our mercy: environmental sins of the father can indeed be passed down to the sons (in the gendered terms of Exodus 34:7). But we can also choose to grace our inheritors with the benefits of righteous stewardship. Consideration of our inheritors entails not only protective actions, but a shift in mentality from the presentism and individualism that marks consumer-driven contemporary life to an awareness of the communal that comes from understanding ourselves in historical terms. This shift of perspective, focusing anew on our place as historical agents, is aided by the concept of generations. As people of the Christian faith, we draw guidance and inspiration from the Bible. The biblical account of generations, however, requires some wrestling on our part, since the two parts of the Christian Bible—the Old Testament and the New (or, the Hebrew and Christian Scriptures)—diverge so strikingly in their depiction of generations. Whereas the Hebrew portrayal of generations is largely positive, the portrayal of generations in the Gospels and Epistles is largely derogatory. Taken together, how does the Christian Bible offer a scriptural language for protecting our inheritors, the children of our children's children?

THE GENERATIONS OF SOLOMON'S TEMPLE

There are many points of entry for reflecting on our position within an intergenerational history. One point is to follow the cedars of Lebanon into the Hebrew Scriptures. The Hebrew Bible is forested with cedars, which are mentioned seventy-three times. (Interestingly, there are no appearances of cedar trees in the New Testament.) In the Psalms, for instance, the cedars are a source of respite ("The mountains were covered with its shade, / the mighty cedars with its branches" [Ps 80:10]) and the focus of God's loving care ("The trees of the LORD are watered abundantly, / the cedars of

Lebanon that he planted" [Ps 104:16]). Throughout the Hebrew Scripture, the cedars, and particularly the cedars of Lebanon, are associated with strength. It is therefore no surprise that Solomon chooses them to build the temple. Not only is cedar mentioned repeatedly as Solomon's building material of choice, there are even the surprisingly prosaic detail about how the timber is to be transported (1 Kgs 5:9) and the work shifts of the laborers (1 Kgs 5:14). Cedar wood is emphatically specified as the wood for the inner sanctuary of the Temple:

> He lined the walls of the house on the inside with boards of cedar; from the floor of the house to the rafters of the ceiling, he covered them on the inside with wood; and he covered the floor of the house with boards of cypress. He built twenty cubits of the rear of the house with boards of cedar from the floor to the rafters, and he built this within as an inner sanctuary, as the most holy place. The house, that is, the nave in front of the inner sanctuary, was forty cubits long. The cedar within the house had carvings of gourds and open flowers; all was cedar, no stone was seen. The inner sanctuary he prepared in the innermost part of the house, to set there the ark of the covenant of the LORD. (1 Kgs 6:15–19)

The heart of the Temple—indeed, the very heart of the religion—is encased in the wood of this tree that endures for a thousand years. This is, in Solomon's words, "an exalted house, / a place for [God] to dwell in forever" (1 Kgs 8:13).

The wood of the cedar of Lebanon is perhaps as close as the vegetable world comes to "forever": the wood is not merely strong, but symbolic of eternity. And indeed, this temporal extension reaches to the life of human beings. In a number of ways, the Temple becomes a model of transgenerational humanity, a place that unites the human beings of past, present, and future. On one level, the Temple looks backwards in time. It is a symbolic spatial reconstruction of the Garden of Eden, the first and perfect natural human habitat.[63] The cedar-clad inner sanctuary, Solomon declares, is also "a place for the ark, in which is the covenant of the LORD that he made with our ancestors" (1 Kgs 8:21). But the Temple is also a symbolic space in which ancestors and unborn children meet. Speaking before the altar in the inner sanctuary, Solomon invokes God's promise that there will always be a descendent of David on the throne, "if only your children look to their

---

63. For parallels between the Temple and Eden, see Stager, "Jerusalem and the Garden of Eden."

way, to walk before me as you have walked before me" (1 Kgs 8:25). Having constructed a house for God to dwell in, Solomon identifies the Temple as a place of prayer, repentance and forgiveness that links the people of the present to those of the past: "When your people Israel, having sinned against you, . . . turn again to you, confess your name, pray and plead with you in this house, then hear in heaven, forgive the sin of your people Israel, and bring them again to the land that you gave to their ancestors" (1 Kgs 8:33–34). Solomon concludes by emphasizing generational continuity: "The LORD our God be with us, as he was with our ancestors; may he not leave us or abandon us, but incline our hearts to him, to walk in all his ways, and to keep his commandments, his statutes, and his ordinances, which he commanded our ancestors" (1 Kgs 8: 57–58). Shortly thereafter, God speaks to Solomon, and makes the intergenerational chain even more personal: "As for you, if you will walk before me, as David your father walked, with integrity of heart and uprightness, doing according to all that I have commanded you, and keeping my statutes and my ordinances, then I will establish your royal throne over Israel forever, as I promised your father David," but "If you turn aside from following me, you or your children, and do not keep my commandments and my statutes that I have set before you, but go and serve other gods and worship them, then I will cut Israel off from the land that I have given them" (1 Kgs 9:4–5, 6–7). The Temple is a locus of both the father and the children. (Of course, the children do promptly stray from God's commandments, and the rest of First Kings chronicles the violence and factions that result from the people turning away from God, beginning even with Solomon himself.)

The building of the Temple thus enfolds the natural world—the land, the trees—and generations of people, reaching outward in time. The Temple is a human construction, as the elaborate description of its blueprint and design indicates, but it is also a natural setting. In Solomon's own palace, he builds a large space called the House of the Forest of the Lebanon, with cedar pillars, cedar beams, and a cedar roof (1 Kgs 7:2–3). In the jargon of today's trendy home designs, he brings the outside in, both in the House of the Forest of the Lebanon and in the inner sanctuary. We can return to a passing but important detail of the sanctuary, that the cedar had "carvings of gourds and open flowers," as the wood comes to life in the form of luscious and fertile vegetation (1 Kgs 6:18). The space is befitting of Solomon, who was known as a poet and as a natural philosopher: "He would speak of trees, from the cedar that is in the Lebanon to the hyssop that grows in the wall; he would speak of animals, and birds, and reptiles, and fish" (1 Kgs 4:33). The Temple, Solomon's ultimate artistic and even scientific creation, presents an idealized space of integral wholeness, a place where nature and the human, and human generations, meet in harmony as they reverence the divine. The long-lived cedars both constitute and represent this integration as they physically and symbolically structure the space.

The account of the construction of the Temple in First Kings is full of references to exact spatial measurement, and it is tempting to draw an analogy with the measurement of time through human lives. But generations are not cubits. There are, to be sure, biblical instances where generations are being counted and measured. In the second of the Ten Commandments, God warns of his jealousy and of the consequences to those who violate the command to not worship idols, threatening to "punish children for the iniquity of parents, to the third and the fourth generation of those who reject me" but promising to show "steadfast love to the thousandth generation of those who love me and keep my commandments" (Exod 5, 6; see also Exod 34:7).[64] There is a form of mathematical calculation happening here. But far more common is the general notion of generations across time, as can be heard in a phrase—"throughout generations"—that pulses through Hebrew Scripture like a refrain: "throughout their generations" (Exod 27:21), "throughout your generations" (Exod 29:42, repeated at 30:8, 30:10, 30:31), "for him and for his descendants throughout their generations" (Exod 30:21), "this is a sign between me and you throughout your generations" (Exod 31:13), "throughout their generations, as a perpetual covenant" (Exod 31:16), "throughout all generations to come" (Exodus 40:15). (And so it continues in Leviticus, as "throughout your generations" and its close variations repeat at 3:17, 6:18, 7:36, 10:9, 17:7, 21:17, 22:3, 23:14, 23:21, 23:31, 23:41, 24:3, 25:30. And so into Numbers . . .) This is not a notion of generations as a series of defined human life spans lining up sequentially like trains on a track, but a diffusive, expansive notion of human futurity.

If Old Testament genealogies are traced through the male line, the actual conditions of pregnancy counter-balance, complement, and literally embody this intergenerational continuity. When a woman is pregnant with a daughter, there is a rather miraculous temporal condition. The mother

64. In the New Testament, the clearest example of calculating time through generations is the beginning of Matthew, which traces the history of Israel: "So all the generations from Abraham to David are fourteen generations; and from David to the deportation to Babylon, fourteen generations; and from the deportation to Babylon to the Messiah, fourteen generations" (1:17).

not only carries the unborn child of the next generation, but the mother's ovaries contain other eggs that hold the possibility of other future lives. And since baby girls are born already having within them all of the egg cells that they will ever have, towards the end of the pregnancy the unborn daughter already contains the biological possibility for the generation beyond her. With modern prenatal imaging, a mother can thus see the next generation which already contains, before even emerging into the world, the next generation; mother, child, and grandchildren, at least in their potentiality, are present in one place. This is a type of nonlinear time that we can find in the refracted phrase "throughout generations," a temporality which isn't restricted to sequential time. This intergenerational temporality is the condition of Solomon's Temple.

DISTANCING GENERATIONS

Intergenerational temporality is very different from that of the timeline, the graphic representation of sequential time that came to dominate Western notions of chronology at the end of the eighteenth century, as the technological advancements of the printing press resulted in a proliferation of volumes featuring elaborate graphs dividing history into linear segments.[65] Sequential time has become so normative in modern Western culture that it can be difficult to think our way into other modes of conceptualizing and experiencing time. Eugen Weber reminds us of the constructed social nature of time, and gives some examples of alternative temporal notions and experiences:

> Time and its divisions are social constructs. Chronology, like other "ologies"—astrology, archaeology, sociology, eschatology—serves ulterior ends and reflects realities quite different from abstract measures.
> Herodotus measured time by generations, as the Etruscans did, and by reigns, as the Egyptians and the Mesopotamians did. Polybius measured time by quadrennial Olympiads; . . . At the end of the sixth century, Gregory of Tours began his *History of the Franks* in the Jewish manner, with the creation of the world—a world he expected to last 6,000 years . . . . But most history, like most people until quite recently, ignored abstract chronology. Time was not linear, but multiple, subjective, and specific to particular situations. . . . Thucydides knows about months and years,

65. See Rosenberg and Grafton, *Cartographies of Time*, chapter 4.

but he locates events in time according to tenured priests and rulers, seasons, and memorable events like battles, plagues, and earthquakes. . . . So what we're dealing with is not time, but times that overlie each other.[66]

"[T]imes that overlie each other" was a long-lived conceptual model, one that endured for millennia before the relatively recent arrival of the sequential timeline. Weber's examples are ancient, but deep into the sixteenth century (when people actually started to use "century," a term that comes from a Roman military unit, as a significant temporal division) we find historiographers carrying on the tradition of layered time. In Shakespeare's main source for his history plays, Holinshed's *Chronicles*, the account of each monarch's reign opens by situating the king in a plurality of temporal frames. Richard II, to take one example, began his reign on the twentieth day of June in the year of the world 5344; in the year of our Lord 1377; 310 years after the Norman conquest of England; "about" the thirty-second year of the Holy Roman Emperor Charles IV; in the fourteenth year of the reign of Charles V in France; and in "about" the seventh year of the reign of Robert II of Scotland.[67] This chronological positioning, in which Christian time (here, the year 1377) is just one of many social and eschatological contexts, prioritizes relationality over the sequential order of solar years. This is relative time.

If we think of chronology as an ideology (another "ology" to add to Weber's list), we can see how our way of understanding time is a choice that reflects our cultural values. At one point, for example, Western time was calculated in terms of the reign of Christ, or the "year of our Lord" (Anno Domini, or AD); 1985 meant it was the 1,985th year in this reign. When a higher value came to be placed on cultural pluralism than on Christian heritage, the cultural winds shifted so that AD was determined to be too Christocentric (even though, of course, Christ's birth was the basis for the chronological system), and so while retaining the old numeration of the Anno Domini system, the dates were re-cast as being in the "Common Era" (CE, or, for earlier dates, BCE). The presumed basis of implied commonality is unclear, but there was no doubt a rapid cultural shift in temporal ideology, so that most of my students haven't the slightest idea why we say that it is 2020. This example of a recent shift in temporal thinking reveals both how overt and how hidden the ideology of time can be.

66. Weber, *Apocalypses*, 7–8.
67. Holinshed, "Richard the Second," 6:415.

An implicit ideological value of linear time is that it creates temporal distance and separation. In pre-modern time, historical difference was not of much import—thus we find myriad medieval paintings of Moses in contemporary clothing, to cite just one example. But temporal distance and separation—plotting events as points on a timeline, rather than overlaying them—works in the service of creating distinction and difference. The historical "us" now becomes distinct and different from a historical "them." Today, an artist who chooses to portray Moses in twenty-first-century fashion would be doing so intentionally, the deliberate anachronism most likely having some political motivation. It matters to us, in a way that it did not to medieval people, that we know the inhabitants of the past wore different clothing. And with clear temporal divisions between a contemporary "us" and a historical "them," the "them" of the past is subject to the projections and prejudices, the forms of "othering," that "thems" are prone to. (Many of my students consider people from the past to be more ignorant and less enlightened than people from our own time, for instance.)

Solomon's Temple, then, represents overlapping, multigenerational time; the modern Western world privileges a linear time that makes generations separate and distinct. Which of these is the temporal ideology of the New Testament? The answer is neither. On the one hand, the frequent eschatological orientation of the New Testament renders earthly time moot. As Weber writes, "The New Testament is largely indifferent to chronology. Paul blamed the astrological superstition of those who observed 'days and months and times and years' (Gal 4:10) and discouraged the Colossians from heeding new moons, sabbaths, and allegedly holy days, 'which are a shadow of things to come' (Col 2:16–17). In the perspective of an imminent Second Coming and the passing of the temporal order, worldly time was of little moment. That may be why Paul neither dated letters nor provided the date of historical events."[68]

On the other hand, across the New Testament there is an insistent presentism, as we can see again in the idea of generations. In Hebrew Scriptures, generations are part of a long view, the idea of "throughout": in Lamentations, for instance, we read "But you, O LORD, reign forever; / your throne endures to all generations" (Lam 5:19); in Daniel, "[God's] kingdom is an everlasting kingdom, / and his sovereignty is from generation

---

68. Weber, *Apocalypses*, 9–10. Dr. Crystal Hall notes of Colossians 2:16–17 that "Paul's references here are likely to Roman civic religion and the imperial cult, which had their own ways of constructing time"; manuscript comments February 2019.

to generation" (Dan 4:3). In the New Testament, however, there is a sharp change. "Generation" is no longer used in the context of thinking about multiple generations stretching across space and time (the there-and-then), but primarily to emphasize the sinfulness of the current generation in the here-and-now. The valorizing "throughout" is replaced by a condemnatory "this," as Jesus says: "But to what will I compare *this* generation? It is like children sitting in the marketplaces and calling to one another" (Matt 11:16, see also Luke 7:31; my italics here and following); "The people of Nineveh will rise up at the judgment with *this* generation and condemn it" (Matt 12:41; see also Luke 11:32); "The queen of the South will rise up at the judgment with *this* generation and condemn it" (Matt 12:42; see also Luke 11:31); "*this* evil generation" (Matt 12:45); "*this* adulterous and sinful generation" (Mark 8:38); "*This* generation is an evil generation" (Luke 11:29).

"This" generation carries accrued guilt: "so that *this* generation may be charged with the blood of all the prophets shed since the foundation of the world, from the blood of Abel to the blood of Zechariah, who perished between the altar and the sanctuary. Yes, I tell you, it will be charged against *this* generation" (Luke 11:50–51). "This" generation is the one that will inflict pain upon Christ: "But first he must endure much suffering and be rejected by *this* generation" (Luke 17:25). And "this" is the generation that will in turn suffer—as is fitting, given the other repeated condemnations—the catastrophes of the Second Coming: "Truly I tell you, *this* generation will not pass away until all these things have taken place" (Matt 24:34; see also Matt 23:36, Mark 13:30, Luke 21:32). Repeatedly, an angry or exasperated Jesus lashes out at the current generation: "You faithless and perverse generation, how much longer must I be with you? How much longer must I put up with you?" (Matt 17:17); "And he sighed deeply in his spirit and said, 'Why does *this* generation ask for a sign? Truly I tell you, no sign will be given to *this* generation'" (Mark 8:12). And, most directly, "You faithless generation, how much longer must I be among you? How much longer must I put up with you?" (Mark 9:19). (We should perhaps recognize that for the gospel writers, active around 70–90 CE, it may have seemed that "their" generation brought the calamity of the destruction of the temple in Jerusalem in 70 CE.[69])

The New Testament message for Christians is not that they are part of an enduring generational flow (as with so many of the generational references in the Hebrew Bible), but that they must struggle to differentiate

---

69. I am grateful to Dr. Crystal Hall for this important reminder.

themselves from the current wicked generation: "Do all things without murmuring and arguing, so that you may be blameless and innocent, children of God without blemish in the midst of a crooked and perverse generation, in which you shine like stars in the world" (Phil 2:14–15). In the Acts of the Apostles, Peter quotes Christ when he urges repentance and baptism, specifying that the promise of baptism "'is for you, for your children, and for all who are far away, everyone whom the Lord our God calls to him.' And he . . . exhorted them, saying, 'Save yourselves from this corrupt generation'" (Acts 2:39–40). Interestingly, for Peter the promise of forgiveness seems to be localized in time (it is only for the immediate hearers and their children, not throughout generations like the covenants with Israel), even if it is geographically diffuse ("all who are far away").

As a general pattern, then, the Hebrew Scriptures emphasize God's actions in the world across subsequent waves of human generations—waves that meet and overlap, as waves can do. The New Testament, by contrast, emphasizes the corruption of the present generation and insists on an eschatological future that is outside of the worldly timeframe. When we consider this in the context of the present environmental crisis, it is easy to perceive the corruption of a generation (a corruption in which most of us need to place ourselves, as we ignore or resist the systemic change needed to rapidly reduce collective carbon dioxide emissions). But to step outside of the problem by turning to an eschatological timeframe condemns future earthly generations to a degraded life. (Or, in worst-case scenarios, no life at all.) An adaptive Christianity, while continuing to be grounded in the promise of grace and the pull of the eschaton, needs a vocabulary that broadens our approach to care for the neighbors of the future, our inheritors. And the resources are at hand. Michael S. Northcott has proposed that the theological "vital resources" for the environmental crisis "may be found in an ecological repristination of central features of the Hebrew-Christian tradition, including the Hebrew tradition of created order."[70] Part of this aspect of created order—exhibited in the form and function of Solomon's Temple—is the transgenerational nature of the natural world, where no generation stands distinct from the other and the present, past, and future are intertwined.

---

70. Northcott, *Environment and Christian Ethics*, xiii. Northcott also identifies theological resources in the notion of a trinitarian creator, natural law ethics, and the ethics of classical and Eastern religious thought.

LAYERED TIME, LITURGICAL PRACTICE

At various points in the history of Christianity, this transgenerational nature of the world has been expressed and practiced in different ways. Patristic authors, following the lead of Origen, emphasized the radically typological nature of Scripture, as moments in the long and ongoing story of humanity meet or resonate with each other. People living in the Middle Ages, when so much of the church's theology and infrastructure was predicated on belief in purgatory, formed intergenerational prayer guilds so that those in the perpetual present could pray for the souls of those who had died in the past. Medieval devotional practices have left us the legacy of a rich tapestry of prayers for previous generations, and the people of the past are often woven into hymns and liturgies. What might an emphasis on the mutual reliance, indebtedness, and vulnerability of generations look like in today's world? If medieval devotion looked to past generations, and if sixteenth- and seventeenth-century reformers looked to the future of the eschaton, and if the pietists and rationalists of the eighteenth through twentieth centuries focused on the present, it is now time for us to fold the people of the future into our church life.

We have the vital resources: the liturgy already contains the elements of a complex temporality, and familiar rituals lend themselves to emphasizing the complex ethics of generational interdependence. Steeped in the symbolics of time, the liturgical celebration can be experienced in ways that make us more theologically aware of—and thus more responsive to—our inheritors. In a previous chapter, I noted that actions follow attitudes; the corollary is also true, that attitudes follow actions, and so the collective actions of a congregation in a liturgical setting can also shape attitudes towards the people of the future.[71] The liturgical forms I know best are those of the Episcopal/Anglican Church and its Book of Common Prayer. That said, over the last few decades liturgical reforms across various confessions (Roman Catholic, Lutheran, Methodist, Anglican, Presbyterian, and the Church of South India) have brought liturgies into close alignment,[72] so while I write from the place of my particular church home, I hope these ideas have ecumenical legs.

In a fascinating book on liturgical theology, Nathan G. Jennings thinks about liturgy as a form of pattern recognition. He writes, "Faith,

---

71. I am grateful to Dr. John Hoffmeyer for this observation.

72. See Bradshaw and Johnson, *Eucharistic Liturgies*, chapter 8.

like any form of human perception, is a noetic event wherein patterns are discerned—in this case transcendent, even divine patterns. Theology, like any form of human knowing, brings these perceptions into intelligibility through contemplation."[73] This work of perceiving happens largely through the vehicle and discourse of analogy:

> Given the qualitative difference . . . between the infinite God and finite creation, divine pattern recognition is severely limited in both what it can confirm and in the relative depth of its explanatory accounts. Revelation is necessary for genuine knowledge of the transcendent.
>
> Discourse about revelation, however, is not univocal, and therefore not literal. Rather, it demands *analogical* language as its most appropriate form. Analogy compares two things based upon a correspondence between the two established in a recognizable set of proportions. Discursively, analogy often appears as a kind of metaphor. And although such theological language must remain strongly *metaphorical*, the realities under discussion are not *merely* metaphorical. The comparison, correspondence, and proportionality may indeed be quite real. That is to say, analogy is *ontological*, not merely linguistic or conceptual artifacts of human projection. Analogy is not imagined, or invented. It is discovered.[74]

In other words, to say that the liturgy is an analogy of the divine does not diminish the liturgy by making it "merely" a form of comparison; we accept that we can only ever speak of God through analogy, but this analogy surpasses the trite comparisons of normal human language (e.g., "my love is like a red, red rose") to express and contain God's truth ("the cedars of Lebanon are like the Kingdom of God"). Jennings goes on to discuss a neighboring form of analogy, that of anagogy.[75] Anagogy changes the directionality or point of view vis à vis the metaphor. When we use theological metaphors like the cedars of Lebanon as an analogy for talking about God, we assume the position of trying to look through something to get a glimpse of the divine. Anagogy allows an imaginative reversal in which we step to the other side of the comparison—we cannot, of course, inhabit the divine perspective, but we can posit a divine reality that is expressed to us through the medium of the metaphor. Instead of thinking about how the cedars are like God, we can think of how God is like the cedars. With these

73. Jennings, *Liturgy and Theology*, 3.

74. Jennings, *Liturgy and Theology*, 4.

75. See Jennings, *Liturgy and Theology*, 7–8.

ideas about analogy and anagogy in mind, we can consider the liturgy as a form of pattern recognition, encompassing liturgical manifestations of time (as both representation and experience) that are part of how we come to understand God: time is like God, and God is like time, in that generations and cosmos are folded together.

The Christian liturgy is an act of time, revolving around an act of memory and re-creation. A familiar story is narrated again: "On the night [Christ] was handed over to suffering and death . . . ." The ritual calls attention to a range of temporal frames that can be significantly overlaid. There is the moment of the ritual that is happening in real time before the assembled congregation (present time of the spacetime continuum); there is the moment of Jesus's last Passover supper with his disciples in a room in Jerusalem (past time); there is the biblical moment of the original Passover in Egypt (deep time); there is the moment of the creation of time itself ("you made all things . . . you created them to rejoice in the splendor of your radiance"[76]) (primordial time); there is the current moment of Christ's mystical presence (present time outside of the spacetime continuum); there is the apocalyptic moment of the promised heavenly banquet (future mystical time); and there is the understanding that this promised future is already in the now (the present future). The emblem of this temporal multiplicity is the lamb: the lamb that is slaughtered in Passover sacrifice (Exod 12: 1–13), the paschal lamb of Christ's crucifixion, the cosmic lamb that takes away the sins of the world, and even the actual lamb that is often consumed as part of an Easter feast. The pairing and combining of these moments infuse the liturgy with temporal and mystical depth. The words of the sacrament both "eternalize the sacrifice of Christ in time and make possible our present involvement and participation in it."[77]

The temporal layering that is essential to the mysticism of the Eucharist is woven into the larger liturgical setting through the lectionary, the established calendar of Scripture readings for worship services. The typological pairing of passages from the Old Testament and the New Testament constructs a multidimensional temporality. Elijah, for instance, links to John the Baptist; Melchizedek links to Christ; Christ's radiance at the transfiguration links to Moses's radiant face when he spoke to God on the mountaintop. We call the relationships among these texts "typological," but we should attend to the valences of the term. The word "typology" was

---

76. Episcopal Church, Book of Common Prayer, Eucharistic Prayer D, 372.

77. Miklósházy, Origin & Development of the Christian Liturgy, 277.

a nineteenth-century invention, a term of German theology that quickly migrated into English usage.[78] The term describes a chronological linearity, a sense of "foreshadowing"—a literary term that was also invented around the same time. As such, the Victorian term "typology" stripped away some of the vestiges of medieval, Catholic allegorical reading practices (rich, complex, and sometimes fanciful modes of interpretation), leaving a more sanitized and orderly sense of historical development that corresponded to the linear time favored by the period. "Typology" also tempered the force of prophecy, a notion that was perhaps too zealous or ecstatic for nineteenth-century Anglican tastes.

A consequence of the emergence of the term "typology" was an emphasis on linear narrative chronology, one that generally has only a forward vector: one reads history, or the Bible, as one would a novel, as a book with sequential chapters and a narrative through-line, with particular incidents in the plot giving a foretaste of what is to come. This notion of history and the movement of time is one that is itself culturally determined. To think exclusively through linear time is to miss the possibilities offered by pre- and post-modern multi-directional time. Instead of understanding the paired readings of the lectionary as chapters in a linear Victorian novel, we can think of them through the kinds of texts available to scholars in the early church (for whom, before the third-century emergence of the codex [a bound set of pages] stabilized the order of the biblical texts, "book" meant a scroll, a form that enabled the different books of the Bible to be paired in creative and flexible permutations) or to contemporary filmmakers (for whom the medium of film offers ways of exploring non-linear narrative time—*Memento, Arrival, Magnolia, Harry Potter and the Prisoner of Azkaban*, and the list goes on.[79])

The lectionary, with its gatherings of texts, presents a model of history that can be thought of as pleated or folded time, in the terms offered by the philosopher of science Michel Serres.[80] This layered, pleated historicity is shared, and enacted, in the sacrament of the Eucharist (as well as that of Baptism, if less insistently). The lectionary and the Eucharist allow for both analogy (point A enables us to better understand point B, as Abraham's near sacrifice of Isaac enables us to better speak of the crucifixion) and anagogy (point B enables us to better understand point A, as the crucifixion

78. This paragraph draws from my essay "Always, Already, Again," 267–282.

79. Wikipedia, "List of Nonlinear Narrative Films."

80. Serres and Latour, *Conversations on Science*, 57–59.

enables us to better speak of Abraham's near sacrifice of Isaac). To shift my own metaphors, we can think of this type of temporality as the quantum mechanics of Christian chronology. Time—human time, God's time—does not move in the strict linearity of modernity, but is folded, pleated, crumpled, spiraled, with atoms resonating with each other across galactic distances or appearing at two places at one time.

This understanding of Christian time, a complex multi-dimensional understanding that was developed over centuries, is one that is open to thinking about our inheritors. Thinking about the future is enabled by our thinking of the past; in pleated time, past, present, and future are all encompassed in "history." This, however, is probably not the temporal sensibility that most people bring to the liturgy. One obstacle to opening the liturgy to temporal expansiveness and inclusiveness—folding in the inheritors as well as the dead—is the degree to which our modern lives have become time-bound. On one level, we are more in tune with the precise movement of time than at any point in human existence. Clock-time has become ubiquitous; the precise time (to the hour and minute, anyway) now appears on our phones, our computers, our home appliances, our buildings, our cars, our clocks and our watches. Most often, the time is presented in numeric (digital) form, rather than the circular form of a clock, the sweep of the hands on the dial presenting an analogue of the sundial and indeed of the planetary motion itself. It is now easy—in some ways even unavoidable—to go through a day knowing where we are in time to the exact minute and with a shared knowledge of the exact time, as our devices are regulated by signals sent from satellites with a precision and uniformity unachievable by hand-set clocks and watches.

But this omnipresent sense of time, in a world that for many of us contains increasingly congested roads and calendars, with work hours that are long and days that are short, ironically can cut us off from natural time. Time zones have disconnected clock time from visible solar movement; we have declared that it is noon across a large band of territory, regardless of when the sun is actually directly overhead. (How can it simultaneously be true noon in cities as remote as Annapolis and Indianapolis?) Time has arguably become more of a number than a point in a day, and days and nights are themselves less distinct than they were before the advent of electricity and 24-hour digital access to the planet. Business/industrial/educational/entertainment time has become de-sacralized.

This de-sacralization threatens to deracinate Christianity from its origins. Traditional Christian liturgies were shaped by celestial time: the Liturgy of the Hours was determined by sundials and observation of the stars, and the movement of the heavenly bodies determined fasts, feasts, and the rhythm of prayer.[81] The distinctive feature of the early Christian timescape was the idea of "cycles-within-linearity," according to Patricia Rumsey, as the daily and seasonal observances dictated by the motion of the sun were overlaid with a teleological understanding of history.[82] The cycles-within-linearity model of time held within it two notions that could be in tension with each other. First, there is the idea that time is inherently sacred, emerging as it does from God's creation of the world, of day and night, of the planets in their celestial rotations; second, there is the idea of "time as needing sanctification, an aspect of a fallen world needing redemption."[83] Rumsey contends that both of these ideas of time are present in the New Testament, resulting in a (perhaps exaggerated) binary between Jesus and Paul: Jesus frequently presents a positive world-view, using images of nature and light, but Paul begins to speak of the world as "an alien environment from which Christians must hold themselves aloof."[84] Rumsey notes that "it is particularly during the liturgy that time collapses into eternity; losing its limitations it becomes one with its fullness."[85] This collapsing of time is related to the dual level of the liturgy: "This linking of heaven and earth, of time and eternity through liturgical prayer, understands the universe to consist of two levels. These two levels stand in opposition to each other. The level of earth is that of time, materiality, and passing away. It is only a poor reflection of the realm of heaven, but it shows the way towards it, pointing to heaven as being more important, the place of virtue and holiness . . . . The liturgy joined together in one the praises of the angels and saints in heaven and that of the people of God on earth."[86]

The ideas about liturgical and spiritual practice that Rumsey explores in eighth- and ninth-century Ireland are at once idiosyncratic to that particular place-time and also illustrative of concepts that extend across Christian history. These ideas also suggest diverse ways of thinking about

81. Rumsey, *Sacred Time*, 36–37.

82. Rumsey, *Sacred Time*, 76.

83. Rumsey, *Sacred Time*, 89.

84. Rumsey, *Sacred Time*, 93.

85. Rumsey, *Sacred Time*, 96.

86. Rumsey, *Sacred Time*, 96.

time that can be adapted—and expanded—to meet our material and spiritual present. This requires a recognition of the variety and flexibility of our temporal models. The shades of time in the liturgy—present spacetime, past, deep, primordial, present mystical, future mystical, and present future—lean towards a spectrum of past times. This emphasis on the past is crucial to the notion of the Eucharist as reiterative: repetition, of necessity, relies on that which came before. But the Christian response to the liturgy has been supple, allowing for varied and transforming relationships to the past. Indeed, accommodating diverse notions of time has been part of the work of the liturgy since the origins of Christianity. On the one hand, there is within Christianity a shared sense that "[a]ny organization of the liturgy must inevitably be related in complex ways to the various cosmic cycles that determine the rhythm of existence in time," in the words of Irénée Henri Dalmais.[87] There has not always been agreement on how this should be achieved; Christians have even clashed over whether the liturgy should be organized according to the lunar or the solar cycles of nature.[88] Should Easter, for example, take place according to a lunar calendar as in the Mosaic Passover (the first day of a full moon following the spring equinox) or on the first day of a solar week (the Sun-day, as our own nomenclature still carries remnants of the ancient calendar)? As Dalmais points out, the Christian conception of time and cultural environments don't always harmonize.[89]

The liturgy, as an analogy for the divine life, is always insufficient, as all human analogies are, but this limitation and discordance are also the stuff of human creativity. The liturgy is always in motion, always in a process of adaptation and revision, always reaching for a more perfect means of expressing the divine—and divine time—for the people who experience it. This re-vision—a re-seeing of the possible—was happening in the days of the early church. The Easter feast (not just the season) was once celebrated for fifty days.[90] But by the fifth century, this extended feast no longer fit the needs or thoughts of the people, as Pierre Jounel notes: "the celebration of the entire paschal mystery as an indivisible unity was already being set

---

87. Dalmais, "Introduction," 1.

88. Dalmais, "Introduction," 3–5. For a succinct account of the varieties of measuring time and competing calendars in the ancient world, see Miklósházy, *Origin and Development of the Christian Liturgy*, 145–56.

89. Dalmais, "Introduction," 4.

90. Jounel, "The Easter Cycle," 57.

aside to some extent as the Church responded to the psychological need which the Christian people felt of honoring successively, in the course of these weeks, first the resurrection of the Lord, then his ascension, and finally the sending of the Spirit on the apostles."[91] The interesting phrase here is "psychological need." The ways in which the laity needed to experience divine time, for whatever complex cultural and ecclesiological reasons, had shifted. To give a more recent example, in the mid-twentieth century the Roman Catholic Church revised its mode of celebrating Palm Sunday. The Holy Week ordo of 1955 amplified the processions with psalms, actually returning to medieval tradition; not only should the people participate in the processions, but it should begin outside the church, to recover the sense of passage from one place to another.[92] Then, "The 1970 Missal made its own the regulations set down in 1955. Meanwhile, however, living conditions had changed and some adaptation was necessary. In the cities it is nowadays often impossible to have 'the solemn procession in honor of Christ the King,' which the 1955 ordo had desired. Sometimes political or social conditions are unfavorable. Consequently, the regulations have been made more flexible, while retaining the essentials of the liturgy of this day."[93] This description illustrates both the historical sedimentation and the morph inherent to liturgy. Jounel references earlier medieval traditions, but of course those traditions are themselves replicating the actions of those in the biblical account for Jesus's arrival in Jerusalem the week of his crucifixion. In re-adopting a custom of processing with palms, the liturgy thus layers present, medieval, and biblical past. But this recovery is adaptive and flexible, attuned to the current social realities and conditions. The emphasis on cities here is interesting—the urbanization of the twentieth century, even as it would seem to move the world further from biblical conditions, paradoxically could bring more people into the collective imagining and iteration of Jesus's entry into a major ancient city. And Jounel's "meanwhile" is significant, marking how the culture rapidly transformed even in a short period (from the ordo of 1955 to the Missal of 1970), showing the ongoing need for flexibility in an ever-transforming world.

The liturgy has always been connected to the culture in which it is practiced, even as the liturgy expresses and shares in the transcendent. In *The People's Work: A Social History of the Liturgy*, Frank C. Senn gives an

91. Jounel, "The Easter Cycle," 57–58.

92. Jounel, "The Easter Cycle," 75.

93. Jounel, "The Easter Cycle," 75.

overview of liturgical developments from the origins of Christianity to the present day. He notes how rituals developed in response to material conditions; he comments, for instance, on how the practice of Christian baptism is indebted to the legacy of Roman bathing technology.[94] He also notes how the liturgical calendar was shaped by patterns of social behavior; the liturgies of Holy Week (Palm Sunday, Maundy Thursday, Good Friday) emerged in response to the rise of pilgrimages to Jerusalem in the fourth and fifth centuries.[95] And he notes the shifting forces of aesthetics; Beethoven's *Missa Solemnis* was created as an attempt "to bring the models of the past to the rescue of the present," as the composer immersed himself not in the music of his near contemporaries but in that of his predecessors Palestrina, Handel, J.S. Bach, and C. P. E. Bach.[96] Senn also notes the common practice of incorporating non-official elements into the liturgy. As a contemporary example, he cites the lighting of the Advent wreath, an observance that migrated from the home into the worship service, becoming an occasion for religious education.[97] And in his discussion on medieval religiosity—noting, along with many historians, that "One of the marks of medieval cultural life in general was its special concern for the welfare of the departed"—he observes that the emergence of All Souls' Day (which originated in 998) and soon All Saints' Day became a busy time of cleaning and decorating graves, and that priests offered requiem masses that involved processing through graveyards to bless the graves.[98]

The medieval cult of purgatory provides an example of Christian practice that focused ethical obligation on human beings outside of a contemporary timeframe. While the (largely non-biblical) concept of purgatory and its attendant theology and notions of sin seem bizarre to most of us today, there were particular cultural reasons (such as the plague) and social formations (such as guilds) that made it reasonable to center religious life on past generations. The medieval observances regarding the dead provide a powerful illustration of how the liturgy has assumed different historical orientations at different points in time. The liturgy not only evolves with its specific cultural context, but also engages differently with temporality and

94. Senn, *People's Work*, 32.

95. Senn, *People's Work*, 63.

96. Senn, *People's Work*, 292.

97. Senn, *People's Work*, 3.

98. Senn, *People's Work*, 199, 201. For an example of historical work on medieval notions of the dead, see for instance, Geary, *Living with the Dead*.

historicity. The practice of the liturgy is not time-neutral. Even as its forms reach back millennia, part of the creative flexibility of the liturgy is how it makes historical sense to worshippers in particular historical contexts.

If in the medieval context there were pressing ethical questions about how to care for the dead, in our own context we now have pressing ethical questions about how to care for the people of the future. Thinking about how to respond to the future has some challenges. Most particularly, our inheritors can't be remembered (obviously), since they have not yet been. And given how rapidly the world is shifting, we cannot imagine what they will be like; we don't know what they wear, what they do, what their world looks like. We don't know their connection to us—while we all have ancestors, we won't all have our own children, so the familial logic of attending to our forebears fails. Our inheritors might be of different colors, and live very differently than us; they may maintain our cultural traditions and artistic forms, or they may reject them. Whereas attending to our forebears could be an act of graciously honoring people about whom we knew something (if not individually, then collectively as part of an age), attending to our inheritors requires a radical act of generosity, an extension of ourselves towards unborn people of whom we know nothing.

LITURGIES OF THE FUTURE

This uncertainty about the future makes the need for a future-oriented liturgy at once more pressing and more difficult. But the task is not impossible. Nor is it huge. Since contemplating the people of the future is an act of historical imagination that invites us to recognize our own place as historical agents, and since the liturgy offers an accretion of historical time, we can work with the liturgy as a form that is both historical and living. Within the flexibility of that form, we can incorporate moments that bring the inheritors into the context of the worship service. Here are some possible ways to do this:

Prayers of the People. This is an obvious place to start. In addition to praying for the souls of the departed, we might pray for the souls of the future. A suggestion: "Merciful God, we pray that you may guide our steps so that we are mindful of the inheritors of the earth who will come after us. Help us to work as stewards of our environment not just for our own sake, but so that we may bequeath a nurturing world to those who will follow." Even a small moment in the liturgy like this can call to mind those of the future, and weave the inheritors into the church's mission of charity. In addition to creating new prayers, we can insert the inheritors into prayers currently in the liturgy, such as this from Form II of the Prayers of the People in the Episcopal Book of Common Prayer: "Praise God for those in every generation in whom Christ has been honored . . . . Pray that we may have grace to glorify Christ in our own day, *and that we have the grace to pass along a healthy planet for the well-being of the generations that will follow us*" (italics marking my addition).[99]

Music. We need contemporary composers like Eric Whitacre to compose choral works for the inheritors. For this perhaps we need poets to write poems that offer inspiration for such compositions. We need both magnificent choral pieces and more humble hymns for congregational singing. We have a number of hymns that recognize the dead (e.g. "For all the saints, who from their labors rest"); why not hymns for those of the future?

Sermons. Writing sermons on the day's lectionary readings might not seem to open space for thinking about the people of the future, but as I hope to have suggested already the lectionary's typological pairings offer rich fodder for contemplating our place in the course of human history. We can attend more to where we are in time (God's time, human time) and to the responsibilities and obligations that come with our own temporal placement.

Confession of Sin. When we confess that we have sinned "by what we have done, and by what we have left undone," might we make that a moment to contemplate our environmental impact? The language of sin is not terribly robust in contemporary Episcopal worship, but perhaps we are at a juncture when we should collectively discuss the language and concept of sin. If earlier generations often concentrated the idea of sin on sexual behavior, perhaps the sin of our time—that which we should be intentionally labeling as "sin," thereby giving it theological force, as I discussed in this book's first chapter—is individual and collective environmental degradation. This idea of environmental sin is something that could be suggested from the pulpit, through reading groups, and in adult and youth education.

Special Collections. We of course have a responsibility to the homeless, those impacted by wars or natural disasters, and others in our community who suffer. But organizations like the Episcopal Relief Services could offer a fund for our inheritors as a way to raise consciousness about those who

99. Episcopal Church, Book of Common Prayer, 386.

come after us. Giving a dollar to a person who will live in the future is an odd thing to do, but it makes a direct connection between our actions now and the consequences for future people. Alternatively (or additionally), there could be a carbon offset fund that would establish ways of helping communities become more resilient to climate change, or to compensate for loss of human habitat and means of livelihood. (Support of a carbon offset fund could easily be done at the level of the parish.) While such a fund risks suggesting that wealthy people can buy their way out of the ethical implications of their carbon usage, it would at least make people more aware of the impact of actions that are otherwise thoughtless, such as turning on the ignition of a car.

Incorporating worship elements that call to mind our environmental actions (or non-actions) and call us to care for the inheritors can be seen as a continuation, and a rounding off, of the liturgical reforms of the last half century. These reforms brought forward patristic liturgical practices and emphasized a common Christian community through shared liturgical shapes. As much as we have turned to the past for grounding, we can turn to the future for inspiration and action.

Attending to our inheritors addresses a unique ethical need of our postmodern moment: recognizing how our actions have repercussions across time, and the need to care for the neighbor across temporal as well as spatial lines. To an unprecedented degree, the current fragility of the planet's climate means that our individual actions resonate—often with profound consequences for the human habitat—in distant places and in future times. This situation requires an expanded understanding of what it means to love God and to love neighbor. Expressions of charity within the liturgy hopefully translate into awareness, which translates into action. While the aim of these proposals is to care for the inheritors (especially the global poor, who are most in peril), acknowledging the inheritors can also deepen our own personal spirituality. Recognizing those who will follow us deepens and widens our sense of human community. And, in much the same way as the Episcopal Eucharistic Prayer C calls us to contextualize our place in space (in "the vast expanse of interstellar space, galaxies, suns, the planets in their courses, and this fragile earth, our island home"), prayers for the inheritors prompt us to recognize our place in time. Like our presence in the vast expanse of space, our presence in the vast expanse of time is small. Acknowledging this smallness—our fleetingness, our life as "a mist that appears for a little while and then vanishes" (Jas 4:14)—can activate the many paradoxes and hierarchical reversals of Christianity, where the low is made high. Small agents of change we might be, but that makes the charge, and the charity, all the greater.

4

Individuality
Parts, Wholes, and the Trinitarian Self

Man is all symmetry,
Full of proportions, one limb to another,
And all to all the world besides;
Each part may call the farthest, brother,
For head with foot hath private amity,
And both with moons and tides.

—GEORGE HERBERT[1]

Our relationship with the environment can never be isolated from our relationship with others and with God. Otherwise, it would be nothing more than romantic individualism dressed up in ecological garb, locking us into a stifling immanence.

—POPE FRANCIS[2]

1. Herbert, "Man," in Di Cesare, ed. *George Herbert*, 40.
2. Pope Francis, *Laudato Si'*, 81.

THE PROBLEM OF INDIVIDUALISM

The previous chapter explored how we might think of the people of the future through the notion of generations. Rather than considering generations as clear units following one another, like the cars of a freight train slowly chugging forward on the linear rails of time, we find that the Bible, the liturgy, and the natural world offer us ways to reconceive of time as intergenerational, with people of the past, present, and future folded together. Where the last chapter contemplated the relationship between groups of people (i.e., generations), this chapter examines a different relationship: that of the person and the community.

A common refrain among ethical and environmental thinkers today is a warning about an individualistic mindset. Bill McKibben, a titan in the world of climate activists, asserts that the most important thing an individual can do about climate change is to stop being an individual.[3] Kathryn Tanner, a powerful voice among contemporary theologians, offers a Christian critique of capitalism that cautions against focusing on personal response. While acknowledging that "fundamental Christian values" such as "love for neighbor, respect for the dignity of the human person, and repudiation of envy and greed in relations with others" are obviously important, the strategy of emphasizing these virtues "can easily suggest . . . that Christianity is primarily concerned with personal morals and not in any direct way with the structures and organization of economic life. The way societies are arranged is not Christianity's business [this argument goes]; only the inward dispositions and attitudes of the individuals who populate them. The result is an overly individualistic approach to complicated structural issues."[4] Pope Francis, the head of the global Catholic Church and a voice of environmental and humanitarian conscience, writes of the climate crisis, "Our difficulty in taking up this challenge seriously has much to do with an ethical and cultural decline which has accompanied the deterioration of the environment. Men and women of our postmodern world run the risk of rampant individualism, and many problems of society are connected with today's self-centered culture of instant gratification."[5] One of the problems resulting from an individualistic culture is that human beings

---

3. Reported in Antal, *Climate Church*, 167, and Sawtell, "Stop Being an Individual," among other places.

4. Tanner, *Economy of Grace*, 3.

5. Pope Francis, *Laudato Si'*, 107.

who fall outside the bounds of the immediate present are often rendered irrelevant. The self's desires in the present trump obligations to the human beings of the past and the future. Or, as the political philosopher Avner de-Shalit writes, "The individualistic self . . . is alienated in relation to the past and future, and thus to the environment, living only in the present. It has no sense of belonging to anything larger than its own private sphere, such as the human race, history, an ecosystem, nature, . . . a cultural transfer from one generation to another, or a creative multigenerational effort."[6]

Thus critiques of individualism, in their various forms, become part of thinking through the ethics of futurity. Climate change communicators emphasize the pitfalls of our culture's emphasis on personal, individual actions. The National Network for Ocean and Climate Change Interpretation (NNOCCI), for instance, trains climate advocates to begin conversations from a shared value, like responsible management ("Why does it matter? What's at stake?"); this leads into a story ("Taking practical, common sense steps to address problems facing our environment today is in the best interest of future generations"). In these conversations there is a need to "strategically redirect thinking away from patterns such as: change is natural/ fatalism; eat it while you can; individualism; nature will fix itself; nature works in cycles; solution = recycling." Advocates are cautioned about the hazards of improperly framing the conversation: "Does it lean too heavily toward consumerist thinking by emphasizing only individual-level actions?"[7] In my own conversations with people about climate change, I have been struck at how quickly the conversation does indeed turn to solution = recycling. Beyond a doubt plastic consumption is a massive problem. In 2017, the rate of global plastic water bottle sales surpassed *one million per minute*, 91 percent of which are not recycled and end up in landfills and water;[8] a report by the Ellen MacArthur Foundation found that we are on a trajectory to have more plastic than fish (by weight) in the oceans by 2050.[9] But while the plastic problem and the climate change problem intersect

6. De-Shalit, *Why Posterity Matters*, 129.

7. These quotations are from NNOCCI handouts distributed at a seminar called "Strategic Framing for Climate Change Communicators," Franklin Institute, Philadelphia, February 23, 2018. For textual clarity I have adapted the formatting from the handout form.

8. Laville and Taylor, "A million bottles a minute"; Nace, "Now at a Million Plastic Bottles Per Minute."

9. World Economic Forum, Ellen MacArthur Foundation, "The New Plastics Economy," 17.

(emissions involved in plastic production and incineration contribute to global warming, for instance[10]), these are really two separate problems, requiring different solutions. Recycling—or, better yet, greatly reducing— plastics helps to address the planetary overflow of trash, but it does little to slow or reverse the earth's rising temperature, the reason for climate change. Turning the conversation to plastics thus evades a discussion of the primary cause of global warming, our collective fossil fuel emissions.

I have wondered why my conversations about climate change keep taking this turn to plastics. In part, the popular focus on recycling could be the consequence of an inaccurate understanding of the causes of climate change. A 2019 poll by the Washington Post and the Kaiser Family Foundation confirmed other studies that show a rapid shift in American attitudes towards climate change, with the vast majority of people now accepting that climate change is real, is caused by human activity, and is a big problem—38 percent of the people surveyed even identify climate change as a "crisis."[11] But when asked what they thought were the sources of climate change, 43 percent identified plastic bottles and bags as a "major contributors," 26 percent identified "burning fossil fuels for heat/electric" as a "minor contributor," and 32 percent identified "driving cars and trucks" as a "minor contributor." Many people, then, seem to have things backwards: our cars and homes are a major source of the global warming that is resulting in climate change, not our plastic yoghurt cups.[12]

These statistics suggest that very well-meaning people might be trying to cut down on plastic use (say, at the church coffee hour) as a way to address climate change, all the while unaware that fossil fuel emissions from their vehicles, houses, and other spaces they occupy (even the church building!) are the real source of the crisis. Maybe the focus on recycling stems from the fact that we can see our plastics, while our automobile emissions are largely unseen. I am just old enough to remember a world before the catalytic converter (introduced in the mid-1970s) rendered vehicle tailpipe exhaust largely invisible—the pictures of cars I drew as a child had billowing clouds of smoke coming from their ends, and I distinctly remember

10. See, for instance, Staley, "Link Between Plastic Use and Climate Change."

11. Guskin et al., "Americans Broadly Accept Climate Science." All subsequent statistics in this paragraph are from this source.

12. As of 2017, transportation was responsible for the most emissions (29% of total U.S. greenhouse gas emissions), followed by electricity (28%), industry (22%), commercial and residential (primarily heating) (12%), and agriculture (9%). EPA, "Sources of Greenhouse Gas Emissions."

watching these clouds from the back seat of the family Ford, mesmerized by their magical swirling motion. Of course those emissions were pollution, our little personal smokestacks. Now that we mostly do not see the direct emissions of the internal combustion engine, perhaps what is out of sight is out of mind, ignored or forgotten.

Or perhaps repressed. Increasingly, I have come to see that the conversational turn to plastics is not always a simple case of truly misunderstanding the cause of the earth's rising temperature. Rather, we look to recycling because it is an easy, specific, and intentional action we can take; by contrast, altering our driving habits or home heating is often not possible without a major change of lifestyle, if at all. (Although electric cars powered from renewable electricity would not require drastic lifestyle changes.) More conceptually, we turn to plastics because the nebulous and collective quality of emissions is something that doesn't accord with our language and mindset. Focusing on plastic reifies—that is, gives a physical shape to—an otherwise invisible problem, re-inscribing an abstraction into the tangible world of material consumer culture and into a moral logic of individualism. Each of those plastic bottles in the sea has someone's name on it, ethically speaking. A problem for which there is no direct moral culpability (global CO_2 emissions) is thus recast, with limited effect, into a discourse of individualism and personal responsibility.

Individualism lies at the heart of the American moral code, and this hinders our ability to address collective social ills. "Rampant individualism," in the Pope's phrase, is the root of the interconnected social/economic/spiritual malaise that is the underlying source of the climate crisis. The planet's warming, in the eyes of many, is not a cause but a symptom of a larger crisis, that of consumer individualism. Even well-intentioned people, raised within a capitalist culture that increasingly emphasizes the efficacy of consumer actions over political ones, often turn to the consumerist logic of individualism as a solution to the climate crisis, seeking redress through personal choices: I will recycle more, fly less, eat more plants, eat fewer animals. While all of these actions are valuable—and, collectively, they do put us on a better path for reducing carbon dioxide emissions—the way to slow and even reverse climate change will require corporate (as in *corpus*, the entire social body) actions and policies, as McKibben argues.[13] Thinking about the problem of climate change solely from the viewpoint of an individual consumer is largely an exercise in futility.

13. See McKibben's eloquent essay, "The Question I Get Asked the Most."

That much seems clear, but there is a deeper, more philosophical consideration here revolving around the relationship of the personal to the collective. Global climate change is an utterly collective crisis: the buildup of CO_2 (and other greenhouse gasses) is a shared consequence, with distributed effects. We cannot attribute personal causality and culpability. (For instance, it would be impossible to say, "That carbon dioxide molecule is from the time I drove to the grocery store, that one is from the time we drove to Colorado for a family vacation in the seventies, that's from the time I flew to London—sorry!" or "That carbon dioxide molecule is mine, that one is Uncle Bob's, that one is Mrs. Cho's, that one is from a guy in Kansas, that one from someone who burned coal in the nineteenth century . . . .") And the effects—flooding in Miami, melting in Siberia, hot nights in Philadelphia—have no one specific cause, but are the result of accumulated human actions. Yet we (that is, the American "we") have been born and bred to conceptualize ourselves as individuals, and to perceive our individualism as a moral and civic virtue. Thus, a massive collective problem faces a nation of individualists: on an existential or ontological level, this does not compute. The problem of climate change does not fit our usual codes of ethics and modes of problem-solving.

The environmental challenge and our self-identity, our self-perception, cannot be easily reconciled. Or, to think of it differently, our outlook towards environmental action cannot be cordoned off from our basic understanding of who we are and how we experience the world. What would it even mean to "stop being an individual" for a people that has long valued individualism? And is that even a desired good? While individualism can be associated with an egocentric consumerism, it is also historically ingrained in the American relationship to nature. The New England author Henry David Thoreau, an icon of American environmentalism and transcendental spirituality, famously went off to live by himself at Walden Pond and wrote, "Individuals, like nations, must have suitable broad and natural boundaries, even a considerable neutral ground, between them."[14] On the other coast, the nineteenth-century naturalist John Muir built himself a cabin in Yosemite, where he could live alone and listen to the water. In our cultural myths and national psyche, being attuned to the natural world often means being by yourself. While we can see the individualism cultivated by consumer culture as a vice, we have long seen environmental solitude—and the corresponding emotional and psychological individualism—as a virtue.

14. Thoreau, *Walden*. The sentence appears in the essay "Visitors."

In the context of conversations about climate change, then, the call to turn away from individualism is not a simple one, and potentially rubs against the grain of much that we cherish and our ways of understanding ourselves. Moreover, discussions of individualism run headlong into a basic philosophical and social problem: how do I maintain my individuality while recognizing my membership in a human collective? This is a problem as old as modernity itself. In the 1590s—during a European period generally known to historians as the "early modern," one that witnessed the emergence of capitalism, a heightened emphasis on the individual person, and a religious shift in emphasis towards personal salvation—William Shakespeare wrote a play called *The Comedy of Errors*. The play revolves around the mix-ups that result from mistaken identity when twins who were separated as babies unwittingly end up in the same place. One of the twins (named Antipholus), newly arrived in the town, laments his condition as a person separated from his family:

> He that commends me to mine own content
> Commends me to the thing I cannot get.
> I to the world am like a drop of water
> That in the ocean seeks another drop,
> Who, falling there to find his fellow forth,
> Unseen, inquisitive, confounds himself.
> So I, to find a mother and a brother,
> In quest of them, unhappy, lose myself.[15]

Shakespeare portrays here a fundamental paradox of modern life: we desire connection to others, but the connection we yearn for simultaneously threatens our very identity. The paradox is visually expressed through Antipholus's beautiful yet absurd aquatic imagery. Yes, in the context of greater humanity, we are all just a drop in the ocean—except that, in actuality, there is no such thing as a drop of water in the ocean. The drop that on land or in air has the distinction and shape conferred by the physics of surface tension ceases to be within the ocean; it becomes just *water*. (Can you picture yourself on the shore, pointing into the ocean and saying to your companion, "Look, what a lovely drop"?) Like the drop of water in his extended simile, Antipholus seeks fellows and fellowship, but the cost of this connection is the loss of self.

The problem is brilliantly synthesized by Shakespeare in one word, "confound." We are blessed in the English-speaking world by the wonder

---

15. Greenblatt et al., *The Norton Shakespeare*, Act 1, scene 2, lines 33–40.

of the *Oxford English Dictionary* (*OED* for short), an expansive resource developed over generations and for more than a century. The *OED* does not merely define a word, but chronicles the meanings the word has accrued and lost over time. According to the *OED*, in Shakespeare's age "confound" meant, "To mix up or mingle so that the elements become difficult to distinguish or impossible to separate" ("confound," *v.*, def. 6). The individual drop in the ocean becomes indistinguishable, impossible to separate, from the water. This is why Antipholus cannot be content. He is driven to be part of the collective and feels forlorn without his family, but the search undermines his personal autonomy: "In quest of them, unhappy, lose myself." For a culture that placed a premium on individuality (Shakespeare's period saw the new popularity of the personal portrait—think Mona Lisa—while we similarly revel in the selfie), this loss of self is viewed not as comedy, but as tragedy.

Shakespeare sets this play—which, for all of its silliness and slapstick, offers a meditation on how we negotiate our sense of self and our desire to be in communal relationship—in the city of Ephesus, in ancient times. This is no accident: *The Comedy of Errors* relies on Pauline theology and precepts. Some of the Pauline references are direct and overt, but there is perhaps a more subtle correlation between Antipholus's meditation on the relationship of self to community and Paul's reflection on finding God. Speaking to the Athenians, Paul says,

> The God who made the world and everything in it, he who is Lord of heaven and earth, does not live in shrines made by human hands, nor is he served by human hands, as though he needed anything, since he himself gives to all mortals life and breath and all things. From one ancestor he made all nations to inhabit the whole earth, and he allotted the times of their existence and the boundaries of the places where they would live, so that they would search for God and perhaps grope for him and find him—though indeed he is not far from each one of us. For 'In him we live and move and have our being'; as even some of your own poets have said, 'For we too are his offspring.' (Acts 17:24–28)

Where Antipholus imagines a quest to find mother and brother in terms of the ocean, noting the absurdity of trying to remain a discrete drop, Paul points out the absurdity of thinking that God lives in a distinct shrine, since everything is of God—we live within God, as fish swim in the sea. Like Antipholus's drop of water that seeks another drop, we search for God, foolishly groping for something to identify, something made of human hands, but that is a nonsensical way of seeking God given the ubiquity of the divine. Rather than aiming to localize God, and to localize God in a material way ("made by human hands"), we would do better to lose ourselves in an all-pervasive deity.

But again, this is easier said than done. I generally experience myself as a discrete self. Physically, I am bounded by skin, and my senses remind me continuously of these boundaries. Psychologically, I have my own thoughts, emotions, and history. I am, sometimes acutely, aware that I inhabit a unique point of view, that which Frederick Borsch (channeling Kurt Vonnegut) called our "peephole" on the world.[16] I do not quite know how to stop being a modern individual. Thus while I appreciate the environmentalist's call to step back from individualism, I find it more useful to think about how, *as an individual*, I relate to the collective. I turn again (and not for the last time in this chapter), to the words of Pope Francis: "Today, the analysis of environmental problems cannot be separated from the analysis of human, family, work-related and urban contexts, nor from how individuals relate to themselves, which leads in turn to how they relate to others and to the environment."[17] Facing our individuality, contemplating how we relate to our distinct selves, is part of the ethical work of contemplating our relationship with humanity and the natural world.

These considerations of self/others/environment are inevitably historical. The "others" that comprise the human collective are persons of the past, present, and future—"all the saints," in the words of the hymn. Yet even as humanity is transhistoric, and our little lives a mere drop in the proverbial bucket of human time, the ways in which we understand ourselves and our place in the world are deeply historically contingent. Acts 17:28—"In him we live and move and have our being"—is an oft-cited verse (it is a favorite of Calvin in the magnificent *Institutes of the Christian Religion*) that emphasizes the ubiquity of the divine and insinuates the a-historicity of God, who is like the air we breathe. But in fact the larger passage is about what it means to be situated in history, about how our individual experience is shaped by cultural distinctions. God places people in a specific time and place—"the times of their existence and the boundaries of the places where they would live"—in order that they would search for God. While God is ubiquitous in spacetime, the particulars of the times and places of a given

16. Borsch, *Spirit Searches Everything*, 129. Borsch cites Vonnegut's *Deadeye Dick*.
17. Pope Francis, *Laudato Si'*, 96.

people differ. This cultural difference is precisely why Paul quotes here not from Hebrew Scripture, but from pagan poetry, speaking to the Athenians in their own idiom. People grope for God within the cultural formations of their own time and place.

Our own cultural formations are marked by two concepts that are part of the intellectual basis of modernity: the idea of the discrete individual, and the concept of an inseparable collective. We speak in a political discourse of individual rights, for instance, and we portray the natural and the human world in terms of systems. But these ideas are not, as it were, *ex divino*, an inherent part of the universe. They are human constructions of a certain time and place—our time and place. One way to track the development of ideas is through tracing the history of the words that express those ideas. Just as the words Shakespeare used (like "confounded") held a specific meaning for his audience, the words we use to talk to each other about value and meaning hold their own contemporary significance. Thinking about the words we use to express ideas is thus a means to delve into the ideas themselves. In the pages that follow, we will think about competing terms, like "individual" and "ecosystem," that shape how we conceive of our relationship to ourselves and to the others of the past, present, and future.

Diving into the words we use to speak about the relationship of selves to communities, of parts to wholes, can reveal an underlying antipathy in our modes of understanding the various relationships we inhabit. There is, or there often can be, a tension between understanding ourselves as persons and understanding ourselves as living in a larger system or economy. At the close of this chapter I propose that we turn to trinitarian theology as a way to rest in the paradox that is our existence as both separate and communal creatures. The Trinity is often invoked in Christian ecological writings, but most often in the context of care for creation.[18] Of course nature is important, but I would suggest that meditating on the Trinity, the difficult concept of three persons as one unity, is a way to contemplate the complexities of our human existence. Considering how we can live, move, and have our being within a trinitarian divinity can move us to approach our climate actions with an intentional recognition of the dynamic interplay and reciprocity of individual and collective actions.

18. See, for instance, the section on the Trinity under "Evangelical Theology and Care for the Earth" in Bouma-Prediger, *For the Beauty of the Earth*, 112–14.

INDIVISIBLE, YET DIVIDED: A HISTORY

A funny thing happened to the word "individual" on its way to the eighteenth century. In 1600, the word pretty much meant just what it looks like: in-dividual, that is, un-dividable, that which can't be separated. By 1700, the word had taken a U-turn in its signification, coming to mean the exact opposite of that which can't be divided: "individual" now indicated something or someone who was distinct, discrete, unique, or separate. In short, "individual" switched from meaning an undividable whole to a part divided off from the whole. The word's change of heart is illustrated by the *OED*. As an adjective, "individual" once meant "One in substance or essence; forming an indivisible entity; indivisible" (adj., def. 1). An example of this usage is a reference from 1580, to "each person of the blessed individuall Trinitie" (with the first such use dating all the way back to 1425 or so). The next definition of an adjectival "individual" is "Of two or more people, things, etc.: that cannot be separated; inseparable" (adj., def. 2). Both of these definitions, the dictionary tells us, gained currency in the mid-sixteenth century but petered out in the mid-eighteenth century. When we hit the third definition, we find the emergence of the word's modern meaning: "Of, relating to, or characteristic of a single person, organism, or thing, or one particular member of a class or group" (def. 3.a). One of the earliest examples of this definition comes from Francis Bacon, the so-called founder of the modern scientific method, who observed in *The Advancement of Learning* (1605) that the manners of learned men are "a thing personal and individuall." Closely related to this idea is definition 4.a: "Existing as a separate indivisible entity; numerically one; single, as distinct from others of the same kind; particular." In short order, then, the word "individual" switches from meaning that which "cannot be separated; inseparable" to a "separate indivisible entity," "single."

Whereas the obsolete definition of "individual" as "indivisible" referenced the divine persons of the Trinity, the newfangled meaning of the word was now used theologically to signify human personhood. In 1593, William Rainolds wrote that "Every man consisting of body and soul, should to his human nature have joyned a particular, a singular or individual subsistence, which Theologie calleth a person or personality" (adj. 4.a).[19] The *OED* also cites another example from Bacon for this definition of

---

19. For the sake of clarity, I have silently modernized some spellings within the *OED* historical definitions.

"individual," noting his reference to the "excellency of your individual person." From Bacon's early seventeenth-century usage the examples quickly jump to a Who's Who list of Enlightenment thinkers: John Locke writes of the "individual Man" in his *Essay Concerning Human Understanding* (1690); Edmund Burke speaks of "any individual servant of the company" (1786); and Alexander Hamilton addresses the "Settlement of Accounts between the United and Individual States" (1793). Once "individual" as an adjective flipped its meaning, the noun form quickly arrived on the scene, meaning "A single human being, as distinct from a particular group, or from society in general" (*noun*, def. B.1.a). Here we find Adam Smith, in his *Inquiry of the Wealth of Nations* (1776), observing that "Among the savage nations of hunters and fishers, every individual . . . is . . . employed in useful labour."

The history of the word "individual" is important because it encapsulates the history of Western individualism. The emergence of the modern idea of the person as a distinct unit is a central part of the story of the Enlightenment. The concept of the individual becomes necessary for scientific investigation (Bacon), the philosophy of the social contract (Locke), ideas of human nature (Burke), the organization of government (Hamilton), and free market economics (Smith). Such individualism—the person is separate and distinct from others—gradually replaces an idea, and an ideal, of persons as interconnected.

The older concept of persons as undivided is shown in John Milton's mid-seventeenth-century poem about Creation, *Paradise Lost*. Here, Adam explains to Eve how they are joined:

> . . . to give thee being I lent
> Out of my side to thee, neerest my heart
> Substantial Life, to have thee by my side
> Henceforth an *individual* solace dear.[20]

There are complicated gender dynamics happening at this moment, but Adam's expressed desire for connection is heartfelt—indeed, "nearest my heart" becomes both a physical expression of connection (as the rib used to create Eve was literally near his heart) and an emotive one (he seeks mutual, undividable closeness, "solace"). And this image of the first humans as blissfully interconnected reflects God's mode of being, and God's mode of being with God's creatures. When Milton's God the Father begets the Son, he declares to the throngs of angels, "Under his great Vice-gerent

---

20. Milton, *Paradise Lost* in *Complete Poems*, Book 4, verses 483–86 (my italics). Subsequent references to the poem will be cited parenthetically by book and verses.

Reign abide / United as one *individual* Soul / For ever happy" (5:609–11, my italics). Not only are God the Father and God the Son interconnected, but through the Son all of the angels can live eternally happy as one united soul. Any wicked angel who dares to disobey the Son "breaks union, and that day / Cast out from God and blessed vision, falls / Into utter darkness . . . without end" (5.612–15). The heavenly life is not one of independence, but of *inter*dependence.

In *Paradise Lost*, this interconnection and mutuality—this individuality, to use the archaic form of the word—of the persons of the godhead and of God and the angels is in itself a reflection of a wider cosmic order. Immediately following God the Father's introduction of the Son, the angels break into song and a "mystical dance" (5:620) which resembles the "starrie Spheare / Of Planets" (5:620–21), with "mazes intricate, / Eccentric, intervolv'd, . . . And in thir motions harmonie Divine" (5:622–25). Milton's "intervolve" is the first recorded use of the word, and is perhaps his coinage. From its Latin etymology, it means "To wind or roll up (things) within each other" (*OED*, verb, 1); the word is a delightfully literal expression of our current saying that someone is "wound up" with others. For Milton, the condition of the cosmos is to be intervolved: stars and planets, Father and Son, angels and godhead, humanity and the divine, man and woman, self and other, human and environment—all are wound within each other. "Intervolve" is also the word's noun form ("An act of intervolving; intertwining") but the dictionary, and perhaps our experience, tells us that this is rare. Unlike Milton's God-angels-humans-cosmos, we perhaps do not always—or even often—feel intertwined with the divine, with each other, and with our environment.

And in fact, as much as Milton's use of "individual" exemplifies a meaning that was going out of style, his idea of God was *avant garde*. Milton's God was not traditionally trinitarian; it is clear that the Son, although divine, is not co-eternal with the Father, and the Son is subordinate to the Father. Even more strikingly, at one point in the poem God complains of his loneliness—"[I] who am alone / From all Eternity" (8.405–6). As one eminent Milton scholar observes, "In the latter part of the seventeenth century, Milton's rejection of the Trinity was at the leading edge of the project of divine isolation, which would soon include not only the singular and self-identical God of Unitarianism but also . . . the rationalist God of the deists, who withdraws from creation after setting it in motion, like a clockmaker who winds a clock and sets it on the shelf."[21] Once the scientific, social, personal, political, and economic self becomes solitary by the eighteenth century, so does the divine self. Ideas about God change over time, as does an understanding of human and human-divine society. In 1624, the poet John Donne famously wrote of humanity in ways that convey its individuality—that is, its indivisibility: "No Man is an *Iland*, intire of it selfe; every man is a peece of the *Continent*, a part of the *maine*; . . . Any Man's *death* diminishes *me*, because I am involved in Mankinde."[22] (Donne's "involved" seems to mean the same as Milton's "intervolved.") But in 1719, Daniel Defoe publishes a bestselling book that is widely considered to be the first English novel, *Robinson Crusoe*, the pseudo-autobiography of a man who is indeed shipwrecked on a nearly uninhabited island, a text that explores the idea of an individual in the modern sense of "a single human being." Not only the content, but the chosen forms of these two authors signal a cultural shift: Donne wrote in poetry and sermons meant to be read out loud in communal contexts; Defoe writes novels, a genre of private, solitary reading.

Over the course of roughly a century, then, there was a cultural shift from extolling the virtues of interconnectivity to lionizing the individual ("A single human being, as distinct from a particular group, or from society in general," to re-cite the *OED*). From psychology to spirituality to economics, the individual self—and most particularly that self's desires—are placed at the center of attention. The philosopher Charles Taylor has written compellingly about this momentous modern Western transformation that re-located the basis of social systems, cultural ethos, and personal lived experience from interdependence to the discrete self. (In Taylor's study, *A Secular Age*, "modern" is the period of 1500 to the present, and "Western" is the region of the North Atlantic, essentially Europe and North America.) Taylor describes "the growth and entrenchment of a new self-understanding of our social existence, one which gave an unprecedented primacy to the individual."[23] Taylor names this process "The Great Disembedding."[24] Prior to modernity, people understood and experienced the world through a pervasive assumption of interconnection; this understanding and

---

21. Shore, "Milton's Lonely God," 36. For the rationalist God, Shore cites Abraham Stoll, *Milton and Monotheism* (Pittsburgh: Duquesne, 2009).

22. Donne, *Devotions*, 299.

23. Taylor, *Secular Age*, 146.

24. Taylor, *Secular Age*, chapter 3.

experience was an integral part of religion, which in turn pervaded the conditions of personal, social, political and intellectual life. But social forces, including religious reform, "open[ed] new possibilities of disembedded religion": "seeking a relation to the Divine or the Higher . . . can be carried through by individuals on their own, and/or in new kinds of sociality, unlinked to the established sacred order."[25] The shift to disembedded religion in turn shifts the fundamental notions of society and social connection, as "the 'world' itself would come to be seen as constituted by individuals," and this too transforms the notion of personhood into an "agent who in his ordinary 'worldly' life sees himself as primordially an individual, that is, the human agent of modernity."[26] As an example of this "remaking of behaviour and social forms," Taylor points to those Protestant churches where membership was not conferred purely by birth into the community, but by joining in response to a personal call: "This in turn helped to give force to a conception of society as founded on covenant, and hence as ultimately constituted by the decision of free individuals."[27] This modern turn to the individual not only had spiritual and social repercussions, but profoundly shaped our understanding of morality. Taylor sums up, "our first self-understanding was deeply embedded in society. Our essential identity was as father, son, etc., and member of this tribe. Only later did we come to conceive ourselves as free individuals first. This was not just a revolution in our neutral view of ourselves, but involved a profound change in our moral world, as is always the case with identity shifts."[28]

Taylor's narrative of this progression is written from the point of view of a philosopher more than a historian, but we can note that the period of time in which this revolution of thought largely took place is the seventeenth through mid-eighteenth centuries—that is, the period of time when the modern Euro-American society and nation state were taking shape. This primacy of the political/social/spiritual individual is itself embedded in our cultural DNA, to the point where we can falsely assume the individual as a given, not as a social construct with a deep and complex history. Taylor's project is motivated by a desire to historicize that which is so ingrained in our cultural mindset that we presume that it is ahistorical, just a basic fact of life. He writes that, "individualism has come to seem to

25. Taylor, *Secular Age*, 154.
26. Taylor, *Secular Age*, 155.
27. Taylor, *Secular Age*, 155.
28. Taylor, *Secular Age*, 157.

us just common sense. The mistake of moderns is to take this understanding of the individual so much for granted, that it is taken to be our first-off self-understanding 'naturally.'"[29]

In academic parlance, the notion of the individual has become "naturalized": it is such an established part of our world that we no longer see it or question it. For an example that illustrates the concept of naturalization, we can turn to a joke that David Foster Wallace told at the beginning of a commencement address. A couple of young fish are swimming along when they pass an older fish who says to them, "Morning boys, how's the water?" The two young fish swim on for a bit, and eventually one turns to the other and says, "What the hell is water?"[30] The point of Wallace's brilliant speech, "This is Water," is that we move through much of life not even aware that we are inside of life. The same is true of our mindset of individualism; for the most part, it is the water in which we swim, and we mostly don't perceive or question it. But it is crucial that we learn to see it and pay attention to it—like the older, wiser fish who knows he is swimming in water—because this fundamental individualism affects how we conceptualize ourselves in relationship to the world and to each other.

In America, then, the idea of individualism is so ubiquitous—the cultural condition in which we live and move and have our being—that we do not even see it. Those of us who were raised in an American context absorbed the primacy of individualism as a natural part of our upbringing. Individual identity (a concept currently emphasized by the left) and individual freedom (a concept currently emphasized by the right) are not just a part of our larger social lexicon, they are the very language in which our culture is written. The notion of the individual is at the core of our legal, social, and spiritual lives. That said, in the United States—which, in many ways, emerged from the intellectual transformations of seventeenth-century England—we still have competing ideals of what it means to be individual, and aspects of the former notion of the "undividable" can still inhere, albeit awkwardly, inside of our notions of individualism. On the one hand, the national motto "e pluribus unum" ("out of many, one")—a motto that appears on the nation's Great Seal and, since the wake of the Civil War, inscribed on all coins—exemplifies the notion of indivisibility, a concept that is explicitly a part of the pledge of allegiance that schoolchildren recite at the beginning of the school day (" . . . indivisible, with liberty

29. Taylor, *Secular Age*, 157.
30. Wallace, "This Is Water."

and justice for all"). (The motto originates in the idea of separate states forming one nation, but it also encompasses separate citizens forming one integral national body.) On the other hand, the notion of the rugged individual is part of our national myth of origins, and many of our cultural values emanate from this figure. The expression "God helps those who help themselves," popularized by Benjamin Franklin, has assumed the cultural status of scripture, even though the saying is non-biblical, and even antithetical to Christ's teachings.

In short, we understand ourselves as both individual (in the word's archaic meaning of indivisible), and as individuals (discrete persons). Like trying to force two positively charged magnets together, it is hard to reconcile the individual with the individual. These seemingly irreconcilable ideals—the incompatible virtues of unity and singularity—make it difficult to think about both our relationship with the natural environment and our relationship with human beings of the past, present and future, since the idea that we are discrete and separate persons divides us not just from each other in the global present, but across historic time. Addressing climate change thus goes far beyond conversations about reducing emissions. We need to dig into the very roots of our personal and cultural ways of being.

ECOSYSTEMS

Around the same time that the word "individual" was reversing its meaning, a new word entered the English language: "system." In providing the word's etymology, our friend the *OED* explains that "system" is a word of "multiple origins," partly borrowed from French, partly borrowed from Latin. The first definition of the word is pretty boring: "*noun* 1. An organized or connected group of things." But then things start to get interesting. The word is a term from ancient Greek music theory used to describe a compound interval, which the *OED* helpfully glosses as an interval composed of stacked tetrachords. It is as a musical term that "system" was first used in English around 1580, per the *OED*, and this meaning quickly takes off in the early seventeenth century. For the early moderns, music was a highly structural concept, and advances in the printing press made it possible to visually depict these complex musical structures. See, for instance, a graph from a 1653 English translation of René Descartes's *Musicae compendium* (Figure 3). It is interesting that Descartes's visualization of music is so systematic in its careful explanation of relationships, since in another vein of his philosophy he propelled the move towards individualism with his famous *Cogito ergo sum*—"I think therefore I am," the endpoint of a radical skepticism that asserts the only thing that that can't be doubted is the act of thinking, and therefore one's own separate existence.

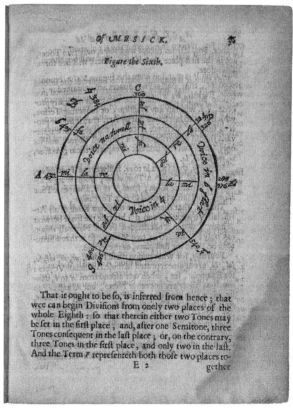

Figure 3. *Renatus Des-Cartes excellent compendium of musick* (London, 1653). **Folger Shakespeare Library.**

Once "system" enters the language, it is used for a host of related meanings: beginning in 1610, it comes to mean "The whole scheme of created things, the universe" (def. 2, now obsolete); as of 1638, "A group or set of related or associated things perceived or thought of as a unity or complex whole" (def. 3.a); in 1651 (by Thomas Hobbes), "A set of persons working together as parts of an interconnecting network" (def. 3.b); in 1665 (by the first English scientific organization, the Royal Society, and in 1690 by John Locke), "A group of natural objects moving in relation to one another under the laws of nature; (*Astronomy*) a group of celestial objects interacting by gravitational forces and moving in orbits about a centre of gravity or central body" (def. 4); and by 1669 the human body itself had systems, "*Biology. A set of organs, other body parts, or tissues having a common structure or function*" (def. 5.a) (as, for example, the nervous system). It would be facile to say that the introduction of the word was the spark that led to an explosion of systematic thinking, but the availability of the word did enable expression for a range of new systems that moved away from older, non-systemic conceptual models (like the Great Chain of Being, or the medical theory of the four humors). If, at the level of the person, things became more individualistic, at the level of so much else—social organization; the cosmos; the body—things became systematic, or at least increasingly perceived as such.

Then, in 1935, a new word was born: "ecosystem." In an article called "The Use and Abuse of Vegetational Concepts and Terms," the British ecologist Alfred George Tansley coined the term "ecosystem" to express a notion of the interrelationship of organisms and their physical environment, with "ecosystem" now "one category of the multitudinous physical systems of the universe, which range from the universe as a whole down to the atom."[31] (The word "ecology," incidentally, only dates from 1866, originally from German; so too *Umwelt*, or "environment" as a particular biological term, comes from German in 1939.[32]) The label and concept of the ecosystem provided a way out of an ecological quandary: how to account for both the part and the whole, or the individual and the indivisible. Earlier biologists had been divided on whether to focus on individuals or the collective. On the one hand, nineteenth-century naturalists (much like the linguistic taxonomists who put together the *Oxford English Dictionary*) were often cataloguers, compiling categories of species and specimens in a way that taxonomically separated individual organisms. On the other hand, there was the idea of holism, in which "physical bodies, chemical compounds, organisms, minds and personalities" constitute biological "wholes."[33] "Ecosystem," which can place an individual organism within a larger context, synthesized the parts and wholes.

31. Arthur G. Tansley, "The Use and Abuse of Vegetational Concepts and Terms," *Ecology* 16.3 (1935) 299. Cited in Golley, *History of the Ecosystem Concept*, 8.

32. Golley attributes the coinage of "ecology" to Ernst Haeckel in *Generelle Morphologie der Organismen* (1866) and "Umwelt" to Hermann Weber in "Zur Fassung und Gliederung eines allgemeinen biologischen Umweltbegriffes, *Die Naturwissenschaften* 27.38 (1939) 633–44; Golley, *History of the Ecosystem Concept*, 2, 42.

33. Golley, *History of the Ecosystem Concept*, 25–26; Golley is describing Jan Smuts's 1926 book *Holism and Evolution*.

"The ecosystem story is largely an American tale," writes Frank Benjamin Golley in his study of the concept. If the idea was born in Europe, World War II derailed European (and Japanese) ecological study. On a practical level there was the fact that many ecologists were dead or elderly, but there was also an ideological reckoning: European "ecologists also repudiated those aspects of ecological theory that had been used by the Nazis and militarists to force conformity on the population and to base racist policy. Ecosystem studies were too close to prewar organismic theories of ecological and social organization to be popular."[34] "In America, however," Golley writes,

> the ecosystem concept appeared to be modern and up to date. It concerned systems, involved information theory, and used computers and modeling. In short, it was machine theory applied to nature. The concept promised an understanding of complex systems and explicitly promised to show how Americans could manage their environment through an understanding of the structure and function of ecological systems and by predicting their responses to disturbance. Further, it extended the holistic concept into the modern, postwar environment. . . . The concept of holism had wider cultural significance. It postulated the existence of a complex entity, larger than humans or human society, which was self-organized and self-regulating. In one sense, the whole was an extension of the Mother Earth idea in modern guise. It involved the extension of God-like or parental properties to nature. Most significantly, it provided the individual faced with the complications and difficulties of daily life the notion that somewhere out there, there was ultimate order, balance, equilibrium, and a rational and logical system of relations. This mixture of ideas was carried forward past the second world war period by the generation that had fought the war, and it dominated the immediate postwar years. The ecosystem concept fit into it, giving guidance to ecological scientists and avoiding dissonance with the overall culture.[35]

The idea of the ecosystem was thus of greater cultural import than the many studies of small lakes in Minnesota would suggest. Ecosystem passes from scientific term to widespread idea to cultural ideology, an ideology that balances more than pollywogs and cattails—it reconciles machine and Mother earth; individual person and nature; the earthly and the divine.

---

34. Golley, *History of the Ecosystem Concept*, 2.
35. Golley, *History of the Ecosystem Concept*, 2–3.

The ecosystem concept also allows for the charting and visualization of abstract or invisible processes. Just as Descartes's diagrams made visible the sonic systems of music, so too diagrams began to give expression to ecosystems. An influential and tragically short-lived early adopter of the ecosystem concept was Raymond Lindeman (1915–42), who included a diagram of a food cycle in Cedar Bog Lake, Minnesota, in each of his major publications. In constructing his model, Lindeman "emphasized the interaction of the living and nonliving parts of the lake, which were intimately interconnected."[36] The image might depict pondweeds, a bog, and plankton predators—I can almost smell the place from here—but it also depicts a sublime natural order and beauty. Indeed, the image itself is strangely beautiful. Around the omphalos of "ooze," there is harmony, balance, dynamic movement yet serenity. The clean lines of the diagram do not depict how the research involved Lindeman taking samples with plankton nets alongside his wife, Eleanor Hall Lindeman—in my imagination, both trawling the one-meter deep lake side-by-side in rubber wading boots and swatting at black flies—but the care and thought of the diagram clearly illustrate that this was a labor of love.[37]

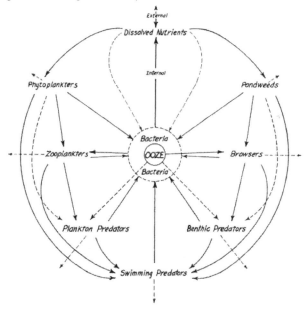

Figure 4. Raymond L. Lindeman, "Seasonal Food-Cycle Dynamics in a Senescent Lake," ***American Midland Naturalist* 26.3 (1941) 637.**

36. Golley, *History of the Ecosystem Concept*, 50.

37. Details of the research drawn from Golley, *History of the Ecosystem Concept*, 49–50.

This irenic vision of the ecosystem meets the reality that as the post-war years turned into the Cold War years, a major sponsor of ecosystem research was the U.S. Atomic Energy Commission, which wanted to know how radioactive materials move through food chains.[38] And the notion of the ecosystem also led to an awareness of the unintended harm of pesticides like DDT, as documented throughout Rachel Carson's pathbreaking and still bestselling *Silent Spring* (1962).

"Ecosystem" did more than enter the dictionary as a word to describe an ecological concept. The word entered the mainstream as a way to express interconnection, the idea that anyone, anything, or any concept is embedded in a complex, interwoven context. The word entered the language as esoteric scientific jargon, but it has since entered popular parlance;[39] it expresses an idea of systemic unity, the idea that an individual is but a part of the whole, and the idea that actions have systemic impact. Words can become worldviews. Scientific principles can become ethical principles. When the Russian scientist Vladimir Vernadsky (1863–1945) published his book *The Biosphere* in 1926 (quickly followed by a 1929 French translation), the book "had wide impact as a scientific expression of a global system of man and nature, which was an antidote to the virulent nationalism that was being expressed at the time."[40] If "individual" reversed itself in the seventeenth century from a term expressing interconnection to one indicating separation, "ecosystem" emerged to convey our indivisibility.

Our notions of ourselves emerge from an ecosystem of ideas, an ecosystem in which concepts large and small influence each other. How we conceptualize our natural environment influences how we conceptualize human lives, and vice versa. This ecosystem is historically dynamic—it shifts over time, prone to both incremental change and periodic shocks that profoundly disrupt flow and relationships. Conversations about climate change are not separate from this ecosystem of ideas, but happen within this larger context. Discussions of policy solutions or technological developments are themselves embedded in intervolved ideas that pertain to the understandings of the self and the community, a community that is variously understood as present and transhistoric.

38. Golley, *History of the Ecosystem Concept*, 3.

39. According to a Google n-gram, which graphs the incidence of word usage for books in Google Books, "ecosystem" starts to show up in the late 1940s, with low usage until the late 1960s, when use of the word spikes to steady usage from the mid-1970s–80s; another dramatic spike in usage occurs in the early 1990s.

40. Golley, *History of the Ecosystem Concept*, 57.

In the ecosystem of ideas, where notions about the human and the non-human interflow, to think about climate is to think more intentionally about ourselves. To think about climate change is potentially to change how we think about ourselves. We have inherited a working vocabulary for expressing ideas, ideas that are themselves shaped—and constrained—by the very words we are given to use. But these words are also labile, prone to transformation and re-definition. We were born into a meaning of "individual," and thus "individualism," that had developed through time. But we are not merely passive recipients of language and ideas. We can actively choose to change meaning and value. Given the complexities of what it means to be an "individual" in an ecosystem, it is perhaps of limited use to the planet and to future generations for us simply to resolve to stop being one. Instead, we can take a more nuanced approach by adapting the concept of the "individual" in a way that reflects our position as both parts and wholes. We can reach into ideas from the past in order to better protect the future. Should we choose not to jettison "individual" but to modulate it in order to reflect the paradox of our separateness and inseparability, we find a theological thought path in the Trinity.

"THE BLESSED INDIVIDUALL TRINITIE"

Words, thoughts, and values thus change with time. But words, thoughts, and values can also carry across time. Perhaps this is a good cultural moment to revive the obsolete meaning of "individual" as indivisible, as that which is ineluctably interconnected. What might happen if, as we hear the word "individual" in our daily lives, we pause to consider that word's origin as a term of trinitarian theology? What might terms like "individual freedom," "individual rights," or "individual responsibility" sound like if we understand "individual" to signify inseparability rather than separate, indivisibility rather than divided? Or, better yet, if we could re-cast "individual" to capture the essence of the both-and, rather than the either-or: we are *both* separate *and* inseparable. Could resurrecting the trinitarian "individual," while folding it into the familiar modern "individual," break through the fantasy of rugged spiritual individualism that is both cause and consequence of consumer capitalism, the logic of which is the underlying source of global climate change?

Trinitarian theology has a rich history that is nearly two millennia deep, with intertwining and divergent strands of thought. One strand places emphasis on the Trinity as divine ontology, an understanding of God's very being as triune, the three-in-one: the Trinity is a description of God's transcendent being. Another strand emphasizes not so much a trinitarian essence as a trinitarian *dynamic* that conveys the flow of divine love between the persons of Father, Son, and Holy Spirit: the Trinity is an expression of participatory divine immanence that extends to human community. (The ancient theological term for this interflow of divine love is *perichoresis*.) Contemporary theology has witnessed a resurgence of interest in this latter perichoretic understanding of the Trinity, led by thinkers such as Jürgen Moltmann. Espousing a "social doctrine" of the Trinity, Moltmann notes that "[t]he monarchical, hierarchical and patriarchal ideas used to legitimate the concept of God are . . . becoming obsolete. 'Communion,' 'fellowship,' is the nature and the purpose of the triune God."[41] Shifting the understanding of God from "monarchical centralism" to a social, perichoretic God has far-reaching repercussions for relationship of various kinds. A triune God "issues an invitation to his community and makes himself the model for a just and livable community in the world of nature and human beings."[42] The Trinity, in this view, teaches a way of living and relating. We look to the triune God as a model or pattern for our personal and communal existence. Moltmann continues, building his argument from a passage beginning at John 17:21, "the character of the primal image in the Trinity does not lie either in the paternal monarchy or the matriarchy of the Spirit, nor in any way does it lie in individual persons, but in the relationships of fellowship between the persons. The level of relations in the Trinity at which the eternal *perichoresis* can be recognized is the element in God from which analogies are to be formed."[43] The Trinity models a relationship of interconnection, in which no one single person (either a person of the godhead or, by analogy, a human person) is distinct from the others.

In a fascinating passage, Moltmann provides an account of the rise of theological individualism that strikingly parallels Taylor's historical narrative of the rise of secular individualism.[44] Moltmann observes that thinking of God in terms of the cosmos "was superseded by the rise of modern, European subjectivity. Once man makes himself the subject of his own world

41. Moltmann, *History and the Triune God*, xii.
42. Moltmann, *History and the Triune God*, xiii.
43. Moltmann, *History and the Triune God*, xvi.
44. Moltmann, *Trinity and the Kingdom*, 13–16.

by the process of knowing it, conquering it and shaping it, the conception of the world as cosmos is destroyed."[45] This is another expression of "The Great Disembedding," with profound consequences for understandings of self, God, and the relationship of the two. "Reality is no longer understood as the divine cosmos, which surrounds and shelters man as his home," Moltmann writes. "It is now seen as providing *the material* for the knowledge and appropriation of the world of man. The centre of this world and its point of reference is the human subject, not a supreme substance. . . . So the unity of what is real is determined anthropologically [i.e., based on the human], no longer cosmologically and theocentrically."[46] Moreover, with the advent of modern European thought that is anthropocentric (i.e., which places human beings at the center of inquiry and value), each individual person comes to determine their own reality, a reality based on one's own experience. Instead of a human life being wrapped in a cosmos pervaded by divinity, it is now the divine that is circumscribed within a person's life: "God is not to be found in the explicable world of things; he has to be sought for in the experienceable world of the individual self."[47] Other theological changes follow from this modern turn: God too becomes thought of "as a subject, with perfect reason and free will, . . . the archetype of the free, reasonable, sovereign person, who has complete disposal over himself."[48] In the nineteenth and twentieth centuries, this understanding became known as "the personal God."[49] In other words, once people began to conceptualize themselves as discrete selves or subjects, they made God in their own image. God became a distinct person, a separate and isolated self. This notion of God is antithetical to a trinitarian, perichoretic God.

A return to a trinitarian understanding of God is therefore not a mere tweaking of doctrine or a theological quibble, but a revolution in thought. Moltmann contends that "there will be a modern discovery of trinitarian thinking when there is at the same time a fundamental change in modern reason—a change from lordship to fellowship, from conquest to participation, from production to receptivity."[50] And this revolution in thought is

---

45. Moltmann, *Trinity and the Kingdom*, 13.

46. Moltmann, *Trinity and the Kingdom*, 13; original italics.

47. Moltmann, *Trinity and the Kingdom*, 15.

48. Moltmann, *Trinity and the Kingdom*, 15.

49. Moltmann, *Trinity and the Kingdom*, 16.

50. Moltmann, *Trinity and the Kingdom*, 9.

not just about ways of understanding God, but a change in thinking that leads to a change in acting:

> We understand the scriptures as the testimony to the history of the Trinity's relations of fellowship, which are open to men and women, and open to the world. This trinitarian hermeneutics leads us to think in terms of relationships and communities; it supersedes the subjective thinking which cannot work without the separation and isolation of its objects.
>
> Here, thinking in relationships and communities is developed out of the doctrine of the Trinity, and is brought to bear on the relation of men and women to God, to other people and to mankind as a whole, as well as on their fellowship with the whole of creation. By taking up pantheistic ideas from the Jewish and the Christian traditions, we shall try to think *ecologically* about God, man and the world in their relationships and indwellings. In this way it is not merely the Christian *doctrine* of the Trinity that we are trying to work out anew; our aim is to develop and practise trinitarian *thinking* as well.[51]

And this revolution in thought is not just about relating to God and each other ecologically—perichoretically, we could say—but about extending this understanding across human and divine time:

> The modern culture of subjectivity has long since been in danger of turning into a "culture of narcissism," which makes the self its own prisoner and supplies it merely with self-repetitions and self-confirmations. It is therefore time for Christian theology to break out of this prison of narcissism, and for it to present its "doctrine of faith" as a doctrine of the all-embracing "history of God." This does not mean falling back into objectivistic orthodoxy. What it does mean is that experience of the self has to be integrated into the experience of God, and that experience of God has to be integrated into the trinitarian history of God with the world. God is no longer related to the narrow limits of a fore-given, individual self. On the contrary, the individual self will be discovered in the overriding history of God, and only find its meaning in that context.[52]

Moltmann is not calling for the elimination of a sense of self, for the figurative drop of water to be absorbed, borderless, into the cosmic ocean. His

51. Moltmann, *Trinity and the Kingdom*, 19–20; original italics.

52. Moltmann, *Trinity and the Kingdom*, 5. For "culture of narcissism," Moltmann cites C. Lash, *The Culture of Narcissism: American Life in an Age of Diminishing Expectations* (New York, 1978).

distinction here is with the self that finds meaning and even greater self-knowledge and identity by recognizing relationship with others and with a God who reaches across earthly, cosmic, and eschatological time. The distinction is between the narcissist, who is self-bounded and looking inward and therefore disconnected from historical (divine) time, and the person who is woven into the fabric of creation, or who participates in the dance of perichoretic interconnection, an intervolvement that acknowledges material, historical, and divine interconnection.

Moltmann's language of ecology thus invites us into a mode of theological reflection with old words (*perichoresis*) and the relatively new (*ecology*). Trinitarian theology uses the vocabulary of economy (*oikonomia*, literally household management, but also God's works in the world), and this inspires comparisons to the modern fiscal economy. But maybe we can turn as well to the model of ecological ecosystem. Lindeman's model of Cedar Bog Lake, with its planktons and pondweeds, might initially seem like a poor model for contemplating the mysteries of the Trinity, but his graph aptly depicts a sense of the dynamic flow of interconnection—we could even think of it as modeling perichoresis and an intertwining of the individual and the whole.

Figure 5. Triquetra. Wikimedia Commons.

An ancient symbol of the Trinity that has also witnessed a renewed popularity is that of the triquetra, or Celtic Trinity knot (see Figure 5), an image that was used for early Christian graves and today's tattoos. Instead of picturing this as a static representation of a three-personed God, we can take it as a graphic portrayal of the interflow and movement of trinitarian life. St. Augustine saw the Trinity as the grounding for a number of triads;[53] to these we could add our own triad of God, environment, humanity. These are inseparably united in a type of ecosystem, even as we enter into this relationship as our own person. One critique of the early modeling of ecosystems was that they did not account for change over time,[54] but in the triquetra the interflow of divinity-nature-human is within the circle of eternity. The divine ecosystem accommodates individual personhood and inseparable persons, as well as discrete moments of time and the intertwined past-present-future.

Moltmann is a leading voice of Protestant theology, but his trinitarian vision shares much with Pope Francis's ethos of interconnection. Indeed, if we were to distill the pope's encyclical *Laudato Si': On Care for Our Common Home* into a one-word thesis statement, that word would be "interconnected." This sense of interconnection appears in a variety of contexts. The biblical stories of creation reveal the interrelationship of human beings and the natural world.[55] We are connected to each other cosmically, since "as part of the universe, . . . all of us are linked by unseen bonds and together form a kind of universal family, a sublime communion which fills us with a sacred, affectionate and humble respect."[56] Interconnection encompasses space, time, and the material world: "It cannot be emphasized enough how everything is interconnected. Time and space are not independent of one another, and not even atoms or subatomic particles can be considered in isolation. Just as the different aspects of the planet—physical, chemical and biological—are inter-related, so too living species are part of a network which we will never fully explore and understand."[57] And, crucially for the central argument of this book, generations are not independent of each other; rather, human beings impact each other and are interconnected across time. This interconnection fundamentally establishes "the rights of future generations."[58] Pope Francis presents this idea in forceful, compelling terms:

53. Some of these triads include the lover, the beloved, and love (*On the Trinity* [*De Trinitate*] 9.2.2); mind, love, and knowledge (9.3.3); memory, understanding, and will (10.11.17); prudence, courage, and temperance (14.9.12); memory, understanding, and foresight (14.10.3). Augustine, *On the Trinity*.

54. Golley, *History of the Ecosystem Concept*, 5, 46.

55. The stories "bear witness to a conviction . . . that everything is interconnected, and that genuine care for our own lives and our relationships with nature is inseparable from fraternity, justice and faithfulness to others"; Pope Francis, *Laudato Si'*, 51.

56. Pope Francis, *Laudato Si'*, 62; see also 64.

57. Pope Francis, *Laudato Si'*, 93.

58. Pope Francis, *Laudato Si'*, 75.

> The notion of the common good . . . extends to future genera-
> tions. The global economic crises have made painfully obvious the
> detrimental effects of disregarding our common destiny, which
> cannot exclude those who come after us. We can no longer speak
> of sustainable development apart from intergenerational solidar-
> ity. Once we start to think about the kind of world we are leaving
> to future generations, we look at things differently; we realize that
> the world is a gift which we have freely received and must share
> with others. Since the world has been given to us, we can no lon-
> ger view reality in a purely utilitarian way, in which efficiency and
> productivity are entirely geared to our individual benefit. Inter-
> generational solidarity is not optional, but rather a basic question
> of justice, since the world we have received also belongs to those
> who will follow us.[59]

The ethos here goes beyond concern for our immediate children and
grandchildren. There is a repositioning of self and human community, an
expansive comprehension of humanity. In the same way that a carbon dioxide molecule endures for centuries after the initial fire that fueled a hearth
or a voyage, our actions—particular and collective—endure across time
and affect other lives. No man is an island, to re-quote John Donne. Or,
more precisely, no person's actions are without consequence to the wider
human and natural world.

What would it look like to live our lives through a trinitarian sensibility? The notion of the Trinity powerfully, mysteriously recognizes the
three-in-one, the distinct identities of Father, Son, and Holy Spirit even as
they are one Godhead. The separate natures of the persons are as important
as the threefold integration. So too we are called to live our lives as both
unique individuals and as part of an inextricable wholeness of humanity.
The model of the Trinity invites us to honor and cherish our distinctive
selves, while also resting in the peace and mystery of our collective cosmic
interconnection. The Trinity radically affirms the both/and of our individu-
alism/indivisibility, rather than the limits and alienation that come with an
either/or.

And how does this trinitarian sensibility inform our climate actions?
The answer is simple: the actions needed to mitigate climate change—given
the massive trans-historic, trans-generational, trans-geographic causes of
the problem—must be both/and solutions. We must take *both* individual
steps *and* collective steps. The situation is too vast and too complex for one

59. Pope Francis, *Laudato Si'*, 105–6.

person to impact; we will need collective policies and responses. Yet the collective is comprised of myriad interconnected persons, so at the same time our discrete actions do have both import and impact. We need to lower (and, eventually, eliminate) our carbon emissions as both persons and societies. Trinitarian thinking enables us to conceptualize our individual climate actions—a change in diet, say, or reducing our use of fossil fuels through the car we choose to drive—as not just personal virtues, but as actions connected to social, systemic efforts to decarbonize modern life. Our personal actions are part of the whole, not endpoint solutions. Moreover, these personal actions can be experienced as a spiritual practice that enacts and reminds us of our participation in a divine ecology.

Our outlook towards environmental action cannot be cordoned off from our basic understanding of who we are and how we experience the world. But we can change who we are and how we experience the world. For Christians, trinitarian theology offers a thought path for intentionally reconceptualizing our relationship—*as individuals*—to a united, inseparable collective. A trinitarian ethos does not demand that we stop being individuals. It does not construct a binary choice between individualism and collectivism. Rather, the Trinity—three persons in triunity—models the both/and of being distinct and inseparable. It respects the individual as well as the interconnection. Resting in this paradox, acknowledging our place within the dynamic of *perichoresis*, changes our sense of relationship with our environment and human others. Just as it is difficult to "stop being an individual," it is difficult to "remember our grandchildren" when we are caught up in a culture of narcissism, or when we try to relate to a God we have made so personal that divinity is confined to the spheres of our separate selves. A vibrant ethos of care for our inheritors does not take hold simply by thinking about future human beings. It begins when we reconceptualize our own identity as both individual and undivided from humans and the natural world across God's time.

5

Hoping Against Hope
Despair, Failure, and Resurrection

The waters wear away the stones;
the torrents wash away the soil of the earth;
so you destroy the hope of mortals.

—JOB 14:19

For there is hope for a tree,
if it is cut down, that it will sprout again,
and that its shoots will not cease.

—JOB 14:7

WHAT'S THE STORY?

Throughout this book, we have considered how the phenomenon of climate change compels us to think of Jesus's dictate to love the neighbor in a new light: given the future impact of our environmental actions, love of neighbor must extend to love of the neighbor across time, to those long dead and especially to those not yet born. Once we expand our temporal understanding of the neighbor, we recognize more fully our obligation to act in ways that care for the people of the future, the inheritors. And once we are in the realm of futurity, the theological virtue of hope becomes an ecological virtue, drawing us to practice better stewardship of the natural world. But this is much, much easier said than done. Working to reduce our individual and planetary CO_2 emissions can easily seem like an exercise in futility. Perhaps the greatest source of climate despair is government inaction. While citizens around the world increasingly understand the urgency of addressing climate change, and while populations in some countries suffer daily from the immediate effects of extreme air pollution that contributes to the long-term climate crisis, people in positions of power (influenced by those with entrenched financial interests) have been sluggish or even resistant to implementing climate solutions. In the United States, with the national government at a historic level of dysfunction and impasse, it is easy to feel overwhelmed by the problem. Much of what needs to be done to reduce CO_2 emissions can be accomplished without national policy and legislation, as Michael Bloomberg and Carl Pope argue in their book *Climate of Hope: How Cities, Businesses, and Citizens Can Save the Planet.* But hope can be fragile and it is often irrational.

This chapter's title, "hoping against hope," is taken from Romans 4:18. Paul is pointing to Abraham's steadfast faith in a future in which his descendants will prosper, even though there is no rational basis for this faith: "[Abraham] did not weaken in faith when he considered his own body, which was already as good as dead (for he was about a hundred years old), or when he considered the barrenness of Sarah's womb. No distrust made him waver concerning the promise of God, but he grew strong in his faith as he gave glory to God, being fully convinced that God was able to do what he had promised" (Rom 4:19–21). A climate of hoping against hope means keeping faith—even in the face of mounting scientific evidence that we are approaching the point of irreversible climate change that will forever alter the face of the planet—that we can still act to change the future. But unlike Abraham's trust in God's promise of future generations, we cannot rest in the faith that God has promised us planetary stability, because no such promise was made: if anything, in the expulsion from Eden God set the terms for a hardscrabble relationship with the earth. Thinking through biblical terms, our care for the environment and thus each other is motivated not by a promise but by an implicit command to rule over nature (Gen 1:26, 1:28, and 2:15, "The LORD God took the man and put him in the garden of Eden to till it and keep it"), with all the responsibilities that come with rulership. So the hope that we need to cultivate is not the patient faith of relying on God's promise (as was the case with Abraham), but the active commitment that comes through our own motivation. Shortly after relating Abraham's hope against hope, Paul asserts that justification in faith leads to peace with God, but he also goes on to describe faith-based grit: "suffering produces endurance, and endurance produces character, and character produces hope" (Rom 5:3–4).

Hope can thus be seen as a product of suffering, endurance, and character. But hope also needs a face and a story. Human beings have always needed stories to crystalize and share ideas. The Bible offers stories of environmental loss: the story of the Garden of Eden tells of an idealized world that was lost because human beings deliberately violated the divine order; the story of Noah's Ark and the great flood tells of the consequences of a population turning away from God's ways. As theologians have reached for a story to tell about climate change, Noah's Ark is one that comes readily to hand, but this can perpetuate a sense of futility. For instance, the work of Michael Northcott, who turns to the story of the Ark, is powerful but often devastating to read.[1] And more broadly, environmental activists have struggled to find a shared story that encapsulates our moment. Some have noted that the dominant climate narrative is the story of Icarus from classical Greek mythology, a figure who flew too close to the sun and was destroyed. If this is so, the story only perpetuates despair and hopelessness.

Seeking a story of hope that can support us in our battle against climate change, Christians might replace the story of Icarus with that of the resurrection. But in doing so, we must acknowledge that the resurrection is hardly a facile narrative, a simple image of an inspirational sunrise and a pat assurance that all will be well. While the resurrection story is indeed the foundation of Christian hope, the complexities of the narrative, especially as told in the Gospel of John, explore the ways in which human failure is not simply overcome, but integrated into the resurrection experience. This acknowledgment of human failings as part of a story of hope helps us to articulate the difficulties, the realities, and the promises of working towards, and for, a better future.

---

1. See Northcott, *A Moral Climate*, 71–77.

THE FALL OF ICARUS

Once upon a time, there was a brilliant inventor named Daedalus. Feeling imprisoned on his island home, Daedalus sought to escape the island by building wings for himself and his son, Icarus. Using all of his innovation, creativity, and technological knowledge, Daedalus skillfully crafted wings out of feathers and wax. With this new invention strapped to their backs, the father and son stood side-by-side, perched to take flight. But first Daedalus cautioned Icarus to keep to a sensible middle path, flying neither too low to the water nor too high towards the sun. As they took flight, Icarus was awed by the magnificence of the view, exhilarated by the sensation of soaring ever upward, and thrilled by the feeling of power in his wings. Disregarding his father's warning, he rose higher and higher, transported by the rush of balmy air on his face, reaching his arms towards the force of the sun. But as he ascended, the heat of the sun began to melt the wax of his wings, and the feathers, molten, dripped into the sea. Icarus hung in the heavens for a moment, as if suspended, then fell and fell to his death in the waters below. The end.

At a recent conference of the Citizens' Climate Lobby, there was a presentation on the stories we tell about climate change. The speaker noted that we are stuck in the story of Icarus.[2] In brief, human beings developed technology, ignored warnings about the environment, and overreached what the planet could bear. Humanity is now being burned by fire and plunging to our death in the ever-rising seas. Our fall is unstoppable. We are doomed. The end. The speaker mentioned that we need to somehow find a different story to tell. For those of us in the Christian tradition, there is another narrative structure ready at hand in the story of the resurrection. But before we think about resurrection, let us take a moment to dwell on Icarus.

A favorite figure of Renaissance authors and painters, Icarus is a character of profound tragedy, with a clear case of what Aristotle called a "tragic flaw" (in Icarus's case, the flaw of hubris). This tragedy is vividly depicted in a painting by the Flemish painter Jacob Peter Gowy (see Figure 6). On Icarus's face we see anguish and fear, one hand in position to brace for a fall, the other in a helpless inverted pose of beseeching prayer. Both will prove futile. We cannot clearly see the expression on the father's face, but the open mouth, the angle of the eyebrows, and the raised furrows in the forehead

2. Lesley Beatty, speaking on the panel "Communications for Climate Advocates," Citizens' Climate Lobby 2018 Congressional Education Day, November 12, 2018.

suggest sheer horror. In the midst of the tragic scene, we are struck by the beauty of Daedalus's wings, redolent of an angel, although the strings and visible handles reveal the construction of the mechanism, rendering the artistry a mocking facsimile of the divine. The victory of human ingenuity proves false, and all is loss, terror, and grief.

**Figure 6. Jacob Peter Gowy, *The Flight of Icarus* (1635–1637).
Museo del Prado, Madrid.**

It is fairly easy to see how the story of Icarus serves as an allegory of anthropogenic climate change. Human ingenuity created wondrous machines, we failed to heed warnings about overreach, and now the very sun is burning us up. The Icarus story lurks behind the titles of scientific articles like, "Broad threat to humanity from cumulative climate hazards intensified by greenhouse gas emissions." The journal abstract for this article can be read as tracing the arc of Icarus's catastrophic fall:

> The ongoing emission of greenhouse gases (GHGs) is triggering changes in many climate hazards that can impact humanity. We found traceable evidence for 467 pathways by which human health, water, food, economy, infrastructure and security have

been recently impacted by climate hazards such as warming, heatwaves, precipitation, drought, floods, fires, storms, sea-level rise and changes in natural land cover and ocean chemistry. By 2100, the world's population will be exposed concurrently to the equivalent of the largest magnitude in one of these hazards if emissions are aggressively reduced, or three if they are not, with some tropical coastal areas facing up to six simultaneous hazards. These findings highlight the fact that GHG emissions pose a broad threat to humanity by intensifying multiple hazards to which humanity is vulnerable.[3]

Simply reading abstracts like this feels like the soul falling. It is perhaps no accident that the Irish Climate Analysis and Research Units goes by the acronym ICARUS.[4]

I personally don't have traceable evidence for 467 pathways in which the planet is going to hell. But I have vivid memories of another type of doomed flight. I spent my childhood summers on the southeastern shore of Lake Michigan, which was on a migration path for the monarch butterfly. We frequently saw individual monarchs towards the end of the summer, but once I was in the right place at the right time, and in the late afternoon sun on the white sand beach I witnessed the peak of the monarch migration. Settled on the shore, the butterflies looked like crisp autumn leaves, strangely out of place without any trees, and strangely out of time in the bright summer sun. But when they took flight—it was magical, wondrous, divine. Monarchs are miraculous creatures, famously migrating three thousand miles every year from the far reaches of Canada to their winter home in Mexico. The delicacy of their translucent, stained-glass wings would seem to defy this endurance, and their capacity for navigation is a marvel.

3. Mora et al., "Broad threat to humanity from cumulative climate hazards," 1062–71.

4. See the ICARUS Climate Research Centre, National University of Ireland Maynooth, https://www.maynoothuniversity.ie/icarus. Dr. Rowan Fealy confirms that the name was chosen because, in addition to the affordances of the acronym, it has "a sun connotation, which is the primary driver of the earth atmosphere system"; e-mail, November 21, 2018.

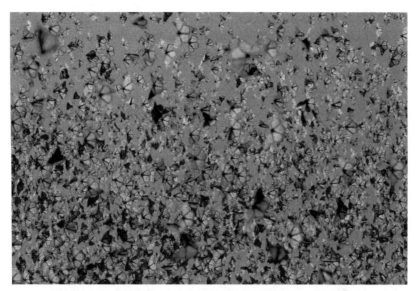

Figure 7. Cloud of Monarch Butterflies. ©Ingo Arndt/naturepl.com.

But a cycle of nature that has continued for millennia is now being threatened by human activity: loss of forest in Mexico, loss of milkweed (the only nursery for monarch eggs) in North America, and the effects of climate change.[5] I continue to spend family summer holidays in Michigan, but these days I rarely see a monarch butterfly. The last one my daughter and I saw together was dead, rolling back and forth in the foamy waves at the shore's edge. I picked it up and gently spread it out on the palm of my hand, noting the brokenness of the wing, the patches where the tiny feathers had fallen away, a miniature figure of an innocent Icarus. The struggle of the monarchs is reflected in a recent flurry of depressing article titles: "Last Monarch Butterfly Migration?"; "End Of Monarch Butterfly Migration Could Be In Sight: Record-Low Numbers Of Monarchs Reached Mexico This Winter, Causing Existential Concern"; "The Last Butterfly"; "Are We Watching the End of the Monarch Butterfly?"[6] Tragically, the question marks in these titles, those little squiggles that offer a glimmer of hope, are increasingly being changed to periods: we are watching the end of the monarch butterfly. In the town of Bolinas (near San Francisco), long a stop of the butterflies' western migratory path, 2017 saw a very low tally of about

5. National Wildlife Federation, "Monarchs Face New Threats."

6. Krahenbuhl, "Last Butterfly Migration?"; Druke, "End of Monarch Butterfly Migration"; Renkl, "The Last Butterfly"; Hannibal, "Watching the End?"

12,360 monarchs; 2018 saw only around 1,250.[7] (On top of the other challenges they face, in 2018 the butterflies also had to contend with massive wildfires, widely considered by scientists to be a consequence of climate change.) In the 1980s, the total population of West Coast monarchs was estimated at 4.5 million; by last count, there were about 28,500, "dipping below the number scientists estimate is needed to keep the population going. This drastic decline indicates the migration is collapsing."[8] The article chronicling this demise doesn't contain the dreaded word in the title, but the word appears in the article's URL: "extinction."

A doctoral student studying monarch populations describes her reaction to the decrease as one of "sorrow."[9] I would describe my reaction to the decline—and to the threatened or impending extinction—of these butterflies that colored the world and the imagination of my childhood more extensively: sorrow, mourning, lament, profound and devastating loss. An overreaction? Perhaps. In an era of rapid environmental destruction and alteration, we are wiping out species every day, and this rate of extinction will accelerate as the planet warms. Monarch butterflies are but one species that could be named in the death rolls. But for me, the monarchs are one of the sacred manifestations of the divine that I cannot pass on to my children. I tried to preserve the dead monarch that I found with my daughter on the beach, so she could appreciate and love what I love. It is carefully laid on cotton in a little box, but the color has faded and it is turning to dust, a drowned-now-desiccated relic of one of nature's glories. I can describe the monarch migrations of my youth, and I can show her pictures of monarchs in their splendor (and yes, I can and have planted milkweed), but all my daughter can experience is the dingy, tattered remains of broken wings preserved by a mother whose mourning she can see but never feel.

As I began working on this book, I discovered there is an emerging literature for people who write on climate change, because it is just too damn depressing. "Climate grief" is a term that has entered the popular lexicon, and I often suffer from this condition. I sometimes feel like Icarus, free-falling into the wine dark sea. Or, even worse, like Daedalus, watching my children fall into a dark and diminished future.

---

7. Hannibal, "Watching the End?"
8. Hannibal, "Watching the End?"
9. Wade, "Monarch Butterflies in Mexico."

Figure 8. Pieter Bruegel the Elder, *Landscape with the Fall of Icarus* (ca. 1558).
Royal Museums of Fine Arts of Belgium. Photo Credit: Scala/Art Resource, NY.

There is another image of Icarus that serves as a different kind of emblem for our reaction to climate change. It is a painting called "Landscape with the Fall of Icarus," long thought to be by the sixteenth-century Dutch painter Pieter Bruegel the Elder.[10] Have a good look at Figure 8. Don't see Icarus, the legendary tragic figure, plunging into the sea? Look to the lower right-hand corner, and you can see the pathetic little legs, giving a last futile kick as Icarus enters his watery grave. Gowy's painting that we looked at earlier places the viewer mid-air, rendering us a levitating close third-party witness to—or even participant in—the desperate moment, privy to the intimate anguish of father and son and the nearly cosmic scale of that human tragedy. By stark contrast, in "Landscape with the Fall of Icarus," as its title suggests, Icarus's fall is marginal and incidental. Within the picture, absolutely nobody seems to notice that Icarus has plunged from the heavens, mostly because they are busy. The sailors on the ship (perhaps

10. An academically sourced Wikipedia entry on the painting notes, "It was long thought to be by the leading painter of Dutch and Flemish Renaissance painting, Pieter Bruegel the Elder. However, following technical examinations in 1996 of the painting hanging in the Brussels museum, that attribution is regarded as very doubtful, and the painting, perhaps painted in the 1560s, is now usually seen as a good early copy by an unknown artist of Bruegel's lost original, perhaps from about 1558." See Wikipedia, "Landscape with the Fall of Icarus."

a symbol of commerce) are doing their own tasks: one climbs the ropes, looking upwards but so concentrated on his own ascent that he seems not to have even noticed a human body falling from the sky; one shimmies along a cross mast to gather up a sail, careful not to fall himself; one has his back turned, bent over as he works on his own task, oblivious to the tragic drama happening behind him. The ship sails steadily onward as part of the nautical traffic and trade. In the center of the painting stands a shepherd, although not a very good one by the look of it. The shepherd seems to gaze aimlessly in the distance; his dog stares elsewhere, so nothing in particular has caught their mutual attention. The shepherd leans casually on his staff; he is not actively tending his flock, and some of the sheep are wandering precariously close to a cliff edge. In the common iconography of the time period, it seems fair to assume that the shepherd represents the church. Then there is the scene of agriculture in the foreground. The draught horse plods heavily in his traces, pulling a plough in rough tillage. Our eyes are drawn to the ploughman because the vivid red of his shirt is the only real color contrast in the painting's palette of browns, greens, and grays, but except for the shirt the man bears an almost comical resemblance to his beast. The ploughman's shaggy hair matches the horse's uneven mane; the angle of the head for both man and horse are the same; the overhang of the man's cap corresponds to the horse's blinders; the gray-green of the man's tunic matches the horse's blanket; the color and angle of the man's legs mirrors the step of the horse's front legs, while the bend of the man's heavy shoe on his right leg mimics the shape of the hoof on the horse's back rear leg. The man wields a whip, suggesting his dominance over the beast, but the whip doesn't seem to actually reach the horse: dominion is illusion, as man and animal mutually plod their way in the ongoing need for feed, two kindred beasts of burden.

Those involved in commercial trade and food production, and even those in the church (with the possible exception of one vaguely curious sheep), are oblivious to Icarus's fall. More alarmingly, even Daedalus seems not to have noticed his own son's demise. Soaring above the horse's rump (about forty five degrees to the northeast, as one looks at the picture) is a winged figure, and another that is visually directly above it, although much further away. The higher creature is a bird, but the lower one looks like a person in a hang glider, legs stretched to the back. It seems fair to assume that this is Daedalus, determinedly focused on his journey, not having looked over his shoulder to check on the flight of his son. And, perhaps most crushing of all, in the lower right-hand corner of the painting there is a lone fisherman, leaning forward with outstretched arm as he casts his rod. We might expect that Icarus is within his field of vision, and that the angler might offer aid, or cry for help, or even just cry for shock and grief. But like the ploughman, the fisherman has his head down and clearly doesn't notice so much as a splash. This is no fisher of men, no potential savior, but someone who just hopes to catch his dinner as he drinks from a cup that looks charmingly—or creepily—like a take-out latte.

The butterflies die. The planet burns. The people drive to the mall.

Taken together, then, these two paintings of the fall of Icarus can be perceived as dueling allegories of our response to climate change. Gowy's picture forces us to face, in unsettling and inescapable proximity, the tragic and terrific grief of a changing world. "Landscape with the Fall of Icarus" presents a collective social failure to even acknowledge that there is a problem; everyone is distracted by the stress or mindlessness of the job, or just isn't paying attention, or is focused on securing the next meal. It is horrible to look; it is horrible to look away.

Breugel's "Landscape with the Fall of Icarus" has been used in other historic contexts to express human apathy or ignorance. The poet W. H. Auden, writing about this painting during World War II, reminds us that apathy in the face of crisis is a common human reaction. Describing the scene of Icarus's fall from the sky, Auden depicts a scenario in which figures casually turn away from the event; the ploughman, for instance, might have heard Icarus's cry and watery landing, but the moment did not hold significance for him as he continued to labor.[11] William Carlos Williams's brilliant meditation on "Landscape with the Fall of Icarus" (also the title of his poem) describes the scene from the point of view of nature itself, and unwittingly describes our new climate problems. For nature, which is concerned with its own well-being, there is an insignificant splash: "this was / Icarus drowning," in the chilling final words of the poem.[12] For the natural world, the fall of the human species will not mean much—if anything, the little splash of falling humans in the long life of the planet will be a relief. If monarch butterflies and polar bears and the human species go extinct, nature will continue, adapting—probably even thriving—in a world where a single cunning species has not become an enemy to the very habitat upon which it depends.

11. Auden, *Collected Poems*, 179.

12. Williams, *Collected Poems*, 386.

If the tale of Icarus is our dominant narrative for talking about global climate change, it is easy to see why we need a new story. The various iterations of the narrative—as full-on Greek tragedy, as a morality tale of human apathy and ignorance, as a planetary sigh of relief—only generate hopelessness, despair, and a sad resignation. If we dwell in and on this story, we not only mire ourselves in depression and defeat, we manage to create a self-fulfilling prophecy through inaction. We need to find alternative stories that prompt hope and action.

THE RISE OF CHRIST

Once upon a time, there was a humble carpenter, skilled at crafting things out of wood. His father, knowing that humanity was imprisoned on its planetary home by greed, power, and narcissism, called him to a higher purpose. Attracting a motley crew of followers, from those who gathered fish to those who collected taxes, the son set out on a mission to teach about a different way of life. Instead of greed, he offered generosity; instead of power, he offered hospitality; instead of narcissism, he offered compassion. Instead of isolation, he offered radical empathy. Many people listened, contemplated his words, and were inspired to change their ideas and their lives. But those who lived by greed, power, and narcissism were threatened by the son's message and feared the love that the people gave him. They nailed him to a piece of wood, the grain and scent of which he understood so well. Before he died, he called out in grief to his father, who seemed to have forsaken him. His followers wept and hid in fear, and then disbanded. The end.

That is the Icarus version of Jesus's life. But of course we know that there is more to the story, and a better ending. The central belief of Christianity is a trust in the resurrection. Without the resurrection, there is only a very sad, and sadly too familiar, story of man's inhumanity to man, another account of a cruel and humiliating death by torture. There is also the too familiar story of those with moneyed interests and government connections shutting down voices calling for change. And certainly we see this phenomenon at work in the context of climate change, as the powerful seek to safeguard a lucrative status quo, even at the expense of human health and habitat. There is plenty of room for gloom and doom. Even as stories of devastating climate events become part of the daily news cycle, there are entrenched interests that are preventing the types of adaptation we need to mitigate climate change. For instance, while agriculture is a substantial contributor to global CO_2 emissions, the policies of the U.S. Farm Bureau, which is closely tied to the lucrative insurance industry through crop insurance, discourage CO_2-reducing transformations in farming practices.[13] But there are also many points of hope: the use of coal in the U.S. has been steadily dropping, while major offshore wind projects are getting launched in the Northeast, to point to just one energy development.[14] Old energy sources are being supplanted by new ones. If one follows the climate and energy news these days, we find ourselves in a Janus moment. Janus is the two-faced Roman god of past and future, endings and beginnings, gates and doorways and transitions. Or to choose a different ancient metaphor, we can see our current environmental moment as one governed by the masks of tragedy and comedy—Melpomene, the Greek muse of tragedy, wails in lament over loss (the loss of coastline; the loss of jobs in coal country) next to Thalia, the grinning mask of comedy that laughs at the prospect of a joyful happy ending (reversing the threat of catastrophic climate change; new careers in energy innovation). Melpomene has been busy reciting the story of Icarus; Thalia still needs a script.

People of faith can find these scripts within their religious traditions. Jews and Muslims, Buddhists and Hindus, Native Americans and spiritual-but-not-religious Transcendentalists (among other groups) all have their own spiritual resources for establishing a climate narrative of resiliency and hope. For Christians, the central story of our faith, that of the resurrection of Jesus Christ, offers itself as a narrative of optimism, of light overcoming the darkness. The ultimate victory of life and light is fundamental to the Christian origin story: "In the beginning was the Word, and the Word was with God, and the Word was God. He was in the beginning with God. All things came into being through him, and without him not one thing came into being. What has come into being in him was life, and the life was the light of all people. The light shines in the darkness, and the darkness did not overcome it" (John 1:1–5). Belief in this light that cannot be extinguished by darkness is the heart of Christian faith.

Darkness and light are also, in narrative terms, structuring elements of the resurrection narrative. Both Mark and Luke relate how, at the time of the crucifixion, the world fell into darkness: at or around noon, "darkness

13. Gustin, "U.S. Taxpayers on the Hook."

14. See, for instance, Popovich, "How Does Your State Make Electricity?"; Dan Gearino, "U.S. Wind Power Is 'Going All Out.'"

came over the whole land until three in the afternoon" (Mark 15:33; Luke 23:44). This moment is a fascinating mixture of the figurative and the literal. This darkness reaches back to the story of Exodus, where God instructed Moses to reach his hand toward heaven "so that there may be darkness over the land of Egypt, a darkness that can be felt"; Moses complied, "and there was dense darkness in all of the land of Egypt for three days" (Exod 10:21–22). The plague of darkness was part of the process of liberation that enabled the Israelites to leave their bondage. For Mark and Luke three days becomes the symbolic three hours of Christ on the cross, but the specificity of the time ("three in the afternoon" in the English translation[15]) also gives the account a degree of quotidian realism, placing the event in the context of the workaday world. There is a similar conflation of spiritual metaphor and realism when, on the third day, the women go to anoint Jesus's body "very early . . . when the sun had risen" (Mark 16:2), or "at early dawn" (Luke 24:1). The timing of the visit to the tomb provides a particular narrative detail to the story: we can imagine the sensations of the early morning, like the distinctive dampness, or the distinctive sounds, or the distinctive bluish light shot through with the day's first cold yet brilliant sunray. But the specificity of dawn also has obvious theological import: the resurrection is marked by the emergence of light from darkness. This is in contrast to, say, the darkest biblical book, that of the relentlessly suffering Job, where darkness is mentioned thirty-nine times (for instance, "If only I could vanish in darkness, and thick darkness would cover my face!" [Job 23:17]) and there is no hope of light. The crucifixion story takes us into that thick darkness, but the resurrection leads us back into the dawning light. It is an ancient trope that threatens to become cliché, but that promise of dawn is the foundation of Christian faith.

That said, images of light in the darkness are not quite a *story*. The actual narrative of the resurrection, especially as it appears in the Gospel of John, is more complex than the dichotomy of light and dark would suggest.

15. There is some ambiguity in the two accounts about the exact time of the crucifixion and death of Jesus. I. Howard Marshall, in his commentary on Luke 23:26–49, notes that noon was the sixth hour in Roman and Jewish time reckoning (with what is translated in English as "three in the afternoon" originally signifying the ninth hour), and accepts the timeframe; Alan Cole, in his commentary on Mark 15:21–47, on the other hand, comments, "Perhaps because of his Roman background, Mark seems to count hours differently from the Greek way, so that we are not sure at what exact time Jesus was crucified." New Bible Commentary (University and Colleges Christian Fellowship, 1994, 1953; electronic text hypertexted and prepared by Oak Tree Software, Inc., version 2.0), in Accordance Bible Software, version 12.3.4.

If the story of Icarus works as a parable of human responsibility for climate change—illustrating the dire consequences of playing with fire, or the repercussions of technological over-reach—the resurrection story presents less of a direct analogy (although we can hope for life overcoming death, perhaps the return of the coral reefs that have been decimated by warming oceans). But if the resurrection story is less overtly ecological, it nonetheless provides us with a central narrative that shifts our focus from an individual hero to community, a community that can only form through personal and collective forgiveness. If we think of human-caused environmental damage as personal and distributed sin (as I discussed in the first chapter), the resurrection story as told in John's Gospel models, however painfully and poignantly, a way for that sin to transform from despair to hope.

It is intriguing that the gospel stories of the resurrection do not, in fact, depict the moment of Jesus's resurrection from the dead. By contrast, other moments of biblical resuscitation are recounted in vivid detail. For instance, Elijah carries the dead son of a widow into an upper chamber, lays the corpse on a bed, lies down on the bed three times, cries out to God to restore the child's life, and the boy comes back to life (1 Kgs 17:17–23). Even more colorful are the particulars of Elisha's revival of the son of the Shunamite woman. The child goes to his father in the field, and suddenly complains of a headache. A servant is ordered to carry the child back to his mother; the Shunamite woman holds the boy on her lap but he dies in her arms. We are left to imagine the scene of agonizing grief, but the Shunamite woman does not wallow in it. Strong of mind and will, and refusing to accept the death of her only and late-born son, she lays him on the bed, saddles up a donkey, and gallops off on a high-speed chase to find Elisha and bring him back (2 Kgs 4:24–25). Then, "when Elisha came into the house, he saw the child lying dead on his bed. So he went in and closed the door on the two of them, and prayed to the LORD. Then he got up on the bed and lay upon the child, putting his mouth upon his mouth, his eyes upon his eyes, and his hands upon his hands; and while he lay bent over him, the flesh of the child became warm. He got down, walked once to and fro in the room, then got up again and bent over him; the child sneezed seven times, and the child opened his eyes" (2 Kgs 4:32–35). The specificity of the scene is so detailed as to be cinematic.

The most dramatic resurrection story in the New Testament is that of Lazarus of Bethany, the brother of Jesus's friends Mary and Martha. When Jesus arrives on the scene, he learns that Lazarus has already been in the tomb for four days (John 11:17) and that the body has started to decompose, as indicated by the stench (John 11:39). At the home of Mary and Martha, there is a crowd of people who have come to mourn and give their support. Witnessing the grief of Mary and those with her, Jesus himself "was greatly disturbed in spirit and deeply moved," and began to weep (John 11:33, 35). Jesus is taken to the tomb, followed by the crowd of mourners who have now become a crowd of spectators. Having commanded that the stone be rolled away from the entrance of the cave where Lazarus has been laid, Jesus "cried with a loud voice, 'Lazarus, come out!' The dead man came out, his hands and feet bound with strips of cloth, and his face wrapped in a cloth. Jesus said to them, 'Unbind him, and let him go'" (John 11:43–44).

This scene is striking not just for its visual and dramatic content, but for its presentation—both by the author of the gospel, and also, within the text, by Jesus himself—as a type of performance. Indeed, the highly public, dramatic nature of the interlude can be seen as a little troubling. Before Lazarus died, Mary and Martha had sent Jesus an urgent message about their brother's serious illness, but upon receiving the information Jesus says to the disciples, almost nonchalantly, "'This illness does not lead to death; rather it is for God's glory, so that the Son of God may be glorified through it.' Accordingly, though Jesus loved Martha and her sister and Lazarus, after having heard that Lazarus was ill, he stayed two days longer in the place where he was" (John 11:4–6). Jesus thus deliberately delays going to his friends, waiting until he knows that Lazarus is dead so that there will be an occasion for Jesus to display his power and be glorified. Subsequently, Jesus states that he is glad that he missed the death, since it provides an opportunity for him to publicly perform a miracle and thus further prove his status to the disciples. ("For your sake I am glad I was not there, so that you may believe. But let us go to him" [John 11:15].) When he finally arrives at the tomb, Jesus prays to God in front of the crowd, explicitly commenting on the highly public nature of the miracle he is about to perform as a way of securing the people's belief: "And Jesus looked upward and said, 'Father, I thank you for having heard me. I knew [sic] that you always hear me, but I have said this for the sake of the crowd standing here, so that they may believe that you sent me'" (John 11:41–42).

Although this episode (which does not appear in the Synoptic Gospels of Matthew, Mark, and Luke) contains one of the most direct and powerful assertions of Jesus's status as Messiah—"I am the resurrection and the life. Those who believe in me, even though they die, will live, and everyone who lives and believes in me will never die" (John 11:26)—in our cynical age there is a disturbing edge to the Lazarus story. Jesus appears to let his friends suffer needlessly in order to set the stage for a dramatic miracle. And while Jesus is genuinely moved by the grief he encounters, there might be something a bit manipulative about how he is playing to the crowd. His vocal prayer to God, which he acknowledges is unnecessary and purely "for the sake of the crowd," seems like a performance. So, too, the detail that he "cried with a loud voice" to Lazarus seems performative—the amplitude of the voice seems to be in the interest of making the command audible to the gathered crowd, rather than to Lazarus who is just inside the cave. Compare the stagecraft and self-fashioning of John's Gospel to the simplicity of a resurrection narrative in Luke:

> Soon afterwards [Jesus] went to a town called Nain, and his disciples and a large crowd went with him. As he approached the gate of the town, a man who had died was being carried out. He was his mother's only son, and she was a widow; and with her was a large crowd from the town. When the Lord saw her, he had compassion for her and said to her, "Do not weep." Then he came forward and touched the bier, and the bearers stood still. And he said, "Young man, I say to you, rise!" The dead man sat up and began to speak, and Jesus gave him to his mother. (Luke 7:11–15)

Here Jesus also performs a miracle in the public eye, but without the theatrics we find in John. While recognizing the very different perspectives and portrayals of Christ in the two gospels, we might still be puzzled by Jesus's behavior at the raising of Lazarus in John's telling—the intentional denial of his friends' pleas to come help them, the withholding of himself until he knew Lazarus to be dead, and the weeping before the crowd and the performative prayer.

The Jesus we know from the Gospels is not cruel or calculating, so to read the story in that way does not correspond with his character elsewhere. The point of the story, it would seem, is to emphasize the necessity of public witness as a means to come to belief. Jesus's theatrics—and they are theatrics—are not those of cynical political theater, but those of drama at its best: theater as a means for collective witness and participation. Jesus doesn't simply raise Lazarus from the dead; he raises him from the dead in the most public way possible, so that the spectators—and us too, as readers or listeners picturing the story in the mind's eye—share a communal experience of the miracle of life overcoming death. Jesus has orchestrated the scene of Lazarus's resurrection so that it is not a private act (like Elijah or Elisha reviving the young boys), or a compassionate passing deed of a faith healer (as in Luke), but a conscious scripting of social witness and belief: this is what resurrection looks like. And it is highly effective. We are told that many of those present believed in him (John 11:45). Soon after, that "many" has expanded into a "great crowd" who come to Mary and Martha's house when they give a dinner for Jesus and Lazarus: "they came not only because of Jesus but also to see Lazarus, whom he had raised from the dead" (John 12:9). (Of course, this success is also the negative turning point of the narrative, since Jesus's and Lazarus's star qualities are part of what leads the chief priests and the Pharisees to start planning to put both of them to death [John 12:10].)

If John's idiosyncratic crafting of the extended Lazarus resurrection story aims to dramatize a moment of collective witness to a miraculous victory over death, why, in John's telling, is Jesus's own resurrection so shockingly different? John 11 (the raising of Lazarus) and John 20 (Christ's own resurrection) stand in relationship to each other as a photograph and its negative—the same image but with inverted color values. In the story of Lazarus, there is a highly public unveiling of the tomb, as the stone is dramatically removed from the mouth of the cave at Jesus's command; in the story of Jesus, a solitary Mary Magdalene arrives in the dark to discover that the stone before Jesus's tomb is already gone (John 20:1). In the story of Lazarus, a performative Jesus prays before a large attentive stationary audience; in the story of Jesus, the small trio of Mary Magdalene, Simon Peter, and the unnamed disciple whom Jesus loved are in chaotic and uncoordinated motion, running at different speeds to and from the tomb (John 20:2, 4). In the story of Lazarus, there is a moment of narrative climax when Lazarus emerges from the tomb, "his hands and feet bound with strips of cloth, and his face wrapped in a cloth"; in the story of Jesus, there is no moment of revelation and resurrection, just an empty space with the discarded linen wrappings and head cloth (John 20:5, 6). John 11 gives us a small piece of theater with a complex plot and high drama; John 20 gives us a narrative lacuna—there is a hole in the text just as there is an empty tomb in the physical landscape.

In John 11, Jesus goes out of his way to create a spectacle, since seeing leads to believing. In John 20, *not* seeing, and not having a plot, leads to believing. Instead of the dramatic narrative arc of a resurrection story in John 11, Jesus's own resurrection is a narrative non-event, a nothingness followed by a series of largely story-less appearances. Indeed, the absence of story is part of what brings people to belief: "Now Jesus did many other signs in the presence of his disciples, *which are not written in this book.* But these [that is, the encounters with the risen Christ that do appear in the Bible] are written so that you may come to believe that Jesus is the Messiah, the Son of God, and that through believing you may have life in his name" (John 20:30–31, my emphasis). While we are privy to a handful of select stories, these are apparently outweighed by those that have gone unrecorded. The untold stories are paradoxically presented as the greater silent testimony of the truth of the resurrection; we are given just enough of a sense of presence to believe, but the real story is in the narrative absence.

We find, then, a narrative lacuna at the heart of a gospel that has insistently associated Jesus with resurrection: John's story of Jesus's resurrection contains no resurrection story. And thus unlike Elijah's or Elisha's or even Luke's resurrection narratives, Jesus's own resurrection is not a cinematic event. We cannot see it. We cannot retell it.

I was struck by the implications of this narrative absence a couple of years ago when I visited the Church of the Holy Sepulchre in Jerusalem, the powerful cultural and architectural phenomenon that has accreted over the presumed site of Jesus's tomb. After my visit to the church, I excitedly called my family to tell them about the experience, and ended up with my teenage daughter on the line. When I told her of my visit to the church built around Jesus's tomb, she broke her façade of adolescent ennui and was clearly impressed, asking, "Really? Is he still there?" I gave her a moment to consider what was wrong with that sentence, and after a long pause, she said, "Oh, yeah." After more than a decade of Sunday school, church choir, and a couple of years in Catholic school, she can surely recite the story of Jesus's nativity in exact detail (per the gospel mash-up version we've adopted for the Christmas pageant), and she could probably do a good job retelling the story of Holy Week events in which Jesus is a strong central character with memorable lines, but she has no real story or vision of the resurrection.

Culturally, perhaps the closest we come to visualizing the idea of the resurrection is in images of the ascension. Since the ascension is a vivid biblical event (Luke 24:51) in a way that the resurrection is not,[16] there is a long history of images depicting Christ's return to heaven; historically, this is probably how many Christians envisioned a post-resurrection Jesus. (By way of contrast, we

---

16. There is one ancient gospel text, the Codex Bobiensis, that does fill in the missing scene of the resurrection from Mark's Gospel, a description that merges resurrection and ascension; Larsen, *Gospels Before the Book*, 117.

don't have many paintings of Jesus barbecuing fish on the beach, an actual biblical event [John 21:9–10].) What kind of an image does the ascension give us of the risen Christ? A quick survey of ascension paintings suggests that many, probably most, of these images prominently feature Jesus's feet. In art from the early Christian to the medieval to the baroque to the modern eras, witnesses are left seeing Jesus's feet as he floats heavenward. Insofar as the ascension has come to substitute for the absent visuals of the resurrection, it provides an image of the bodily Jesus leaving our earthly midst and returning to heaven as the mystical Christ. It is an image that depicts a growing distance, both in space and in ontology. This distance can be seen in Hans Süss von Kulmbach's *The Ascension of Christ* (1513) (Figure 9) through the distinction between Christ's feet and the old man's feet in the foreground: although they are both still human feet—the commonality heightened by the fact that both Christ's and the old man's robes are the same shade of red—the change between them is becoming apparent as Christ enters a different plane. We can see this even more strikingly in a painting by Salvador Dalí called "Ascension" (1958).[17] Here the humanity of the feet is clear and tangible, but the rest of this body is entering a spiritual space that is beyond human experience and comprehension, a primordial space of seed and egg and sun and Holy Spirit brooding over the waters. Rather than staring at the soles of Jesus's (noticeably unwounded) feet as a sign of shared humanity, the unnatural perspective serves to alienate us from this experience, and signals our physical and spiritual place as somehow below. We are, as it were, being left behind.

In thinking about the resurrection as a narrative of hope, then, we encounter something of a warped progression. In the Gospel of John, we find Jesus raising Lazarus from the dead, but this turns out to be not so much a preview as the antithesis of his own resurrection. John's resurrection story (like those of Mark and Luke[18]) is not that of a triumphant hero, but a series of immediate reactions in "terror and amazement" (Mark 16:8) to the empty tomb, often followed by scared silence or secretive whisperings of what has happened. Lacking a dramatic narrative of the event at the heart of the

17. I was unable to obtain permission to reproduce this image, but the interested reader is encouraged to do a quick internet search for "Salvador Dali Ascension"; the image is available on multiple sites, like the WikiArt site, https://www.wikiart.org/en/salvador-dali/ascension, accessed January 2, 2020. It is definitely worth a look.

18. Matthew's telling has more drama: when the two Marys visit the tomb, "suddenly there was a great earthquake; for an angel of the Lord, descending from heaven, came and rolled back the stone and sat on it. His appearance was like lightning, and his clothing white as snow" (Matt 28:2–3).

Christian faith, many have turned to substituting the visual image of the ascension for the resurrection. But the pictorial tradition of the ascension—the only real moment when we find Jesus Christ as neither entirely human nor divine Word, but in a transitional state—arguably serves to position us at our most ontologically remote from the second person of the Trinity. Also, to be honest, the images of Jesus's feet at the ascension can seem a little goofy, as can some of the images of Jesus floating upward on a cloud. The German word for the ascension, *Christi Himmelfahrt*, originally meant "Christ goes to heaven," but as the verb *fahren* came to mean driving an automobile, the word has morphed into sounding like "Christ drives to heaven," invoking images of the ending of the movie *Grease*, with Danny and Sandy heading off into the clouds in their souped-up red convertible. There is nothing humorous in paintings of the crucifixion or Christ meeting the travelers on the road to Emmaus, but the history of ascension and ascension-as-resurrection images leaves us with a misguided, and sometimes even cheesy, visual memory.

Figure 9. Hans Süss von Kulmbach, *The Ascension of Christ* (1513).
The Metropolitan Museum of Art, Rogers Fund, 1921.

If we are left, then, with a resurrection missing from its own story, and a pictorial tradition that is theologically and culturally tainted, how can the resurrection function as a narrative of hope? The answer lies in us. In the Gospels, the post-resurrection story immediately shoots out horizontally: rather than a dramatic and heroic vertical ascension to heaven, the gospel writings direct the energy outward into the community. Matthew's Gospel ends only fifteen short verses after the angel appears to Mary at the tomb, verses that include Jesus's "sudden" appearance and greeting to the Marys, his meeting with the eleven remaining disciples on a mountain at Galilee, and Jesus's concluding injunction of the book: "Go therefore and make disciples of all nations . . . . And remember, I am with you always, to the end of the age" (Matt 28:19–20). The ending(s) of Mark's Gospel relates a series of Jesus's appearances—to Mary Magdalene, to two others who "were walking into the country" (Mark 16:12), to the eleven disciples as they sat at the table— and culminates in the mandate to "Go into all the world and proclaim the good news to the whole creation" (Mark 16:15).[19] Luke's Gospel similarly ends with a series of Jesus's post-resurrection appearances to the women at the tomb, to the two on the road to Emmaus, and to the disciples and their companions at mealtime. The gospel again draws to a close with a mandate: "repentance and forgiveness of sins is to be proclaimed in his name to all nations, beginning from Jerusalem. You are witnesses of these things" (Luke 24:47–48). What was not actually witnessed—Jesus sitting up and walking out of the tomb, Lazarus-like—becomes less important than the disseminated message about repentance and forgiveness.

FROM FAILURE TO REDEMPTION

Returning to John's Gospel, we find that this message of repentance and forgiveness has particular force. A familiar mantra of creative writing instruction is that the author should show, not tell. The Gospel of John does just this: it does not end with a rousing mandate to spread the good news about repentance, but with a poignant scene of shame and regret being overcome by love and forgiveness. The risen Jesus now appears at the

---

19. It should be noted that the ending of Mark is a supplement gathered together from the other Gospels; see Alan Cole's commentary for Mark 16:9–20. *New Bible Commentary* (University and Colleges Christian Fellowship, 1994, 1953; electronic text hypertexted and prepared by Oak Tree Software, Inc., version 2.0), in Accordance Bible Software, version 12.3.4.

Sea of Galilee. A group of the disciples, including Peter, have gone back to their livelihood of fishing. They have had a depressing and frustrating night in their boats, with only empty nets to show for their labors. Then, "just after daybreak," as the gospel tells us, "Jesus stood on the beach; but the disciples did not know that it was Jesus" (John 21:4). He advises them on how best to cast their nets and they follow his advice, quickly rewarded with an overflowing catch. Then the disciple whom Jesus loved recognizes him, and tells Peter that this stranger on the beach is Jesus. Peter's reaction is startling: "When Simon Peter heard that it was the Lord, he put on some clothes, for he was naked, and jumped into the sea" (John 21:7). This NRSV translation has Peter rather neutrally "jumping," but other translations have him more assertively "plunging" (per the New English Translation), or even drastically "casting himself into the sea" (in the words of the King James Version).[20] And the Greek text *ebalen heauton* (ἔβαλεν ἑαυτὸν) does indeed mean "threw himself." Peter's reaction at the news of Jesus's presence is odd and unexplained, but it does not appear that Peter is overjoyed by Jesus's return. Certainly Peter does not race toward shore to be with him. Why does Peter have this response? He doesn't seem to be acting out of fear of a ghost (why wait to get dressed if one is reacting in terror?), nor does it seem like an instance of spontaneous Victorian modesty from a fisherman who normally works without much clothing.[21] Instead, we find a variation of a familiar scene of guilt:

> Then the eyes of [Adam and Eve] were opened, and they knew that they were naked; and they sewed fig leaves together and made loincloths for themselves. They heard the sound of the LORD God walking in the garden at the time of the evening breeze, and the man and his wife hid themselves from the presence of the LORD God among the trees of the garden. But the LORD God called to the man, and said to him, "Where are you? I heard the sound of you in the garden, and I was afraid, because I was naked; and I hid myself." (Gen 3: 7–10)

---

20. New English Translation, copyright 1996–2005, by Biblical Studies Press, in Accordance Bible Software, version 12.3.2; King James Version with Strong's Numbers (KJVS), public domain, formatted and corrected by OakTree Software, Inc., Version 3.5, in Accordance Bible software, version 12.3.2.

21. It is likely that "naked" (*gumnos* [γυμνός]) here does not mean completely without clothing, but more specifically without outer clothing, which the word *ependutēn* (ἐπενδύτην) in the Greek text actually signifies.

Peter is most likely recalling, with the sting of recent shame, his threefold denial of Jesus in his Lord's time of trial. Peter wants to hide, to disappear behind clothing and in the waters of the sea, so that the friend he betrayed cannot see the guilt, the self-loathing, the embarrassment in his eyes. In Genesis, the angry God who comes across the hiding Adam and Eve launches into a litany of judgments and terrifying curses. But in John's Gospel, Jesus simply says, "Come and have breakfast" (John 21:12). The original Greek text just speaks of dining more generally (*aristēsate* [ἀριστήσατε]), but the NRSV's "Come and have breakfast" is a brilliant English rendering of quiet, calming friendship.

And we find yet another biblical repetition, another form of replaying an earlier scene. When Jesus had been seized by the soldiers and taken to the high priest, Peter was brought into the high priest's courtyard, as out of place as a fisherman on land, or a humble man in a place of power, or a member of a community alone and surrounded by hostile strangers. These strangers in the dark asked him, curiously and aggressively, if he was a disciple of Jesus; terrified, Peter thrice denied that he knew Jesus—his leader, guide, friend, and messiah (John 18: 17, 25, 27). Peter never had the opportunity to speak to Jesus before the crucifixion, was never able to apologize for his cowardice and betrayal. Now, on the beach, he is asked again three times about his love for Jesus.

> When they had finished breakfast, Jesus said to Simon Peter, "Simon son of John, do you love me more than these?" He said to him, "Yes, Lord; you know that I love you." Jesus said to him, "Feed my lambs." A second time he said to him, "Simon son of John, do you love me?" He said to him, "Yes, Lord; you know that I love you." Jesus said to him, "Tend my sheep." He said to him the third time, "Simon son of John, do you love me?" Peter felt hurt because he said to him the third time, "Do you love me?" And he said to him, "Lord, you know everything; you know that I love you." Jesus said to him, "Feed my sheep." (John 21:15–17)

Peter's three denials of his lord are now replaced with three affirmations of his love. At the moment of the denials, Peter was cold, anxious, physically and emotionally exhausted, and vulnerable as he warmed himself at a charcoal fire with menacing strangers (John 18:18). Now he is sitting on a beach at dawn with a full belly and a boat full of fish, surrounded by friends, beside a charcoal fire Jesus had made, "with fish on it, and bread" (John 21:9). Before Jesus speaks to Peter after breakfast, there is not any indication that the two have yet spoken. We can perhaps imagine Peter's awkwardness before the exchange, and (although Peter's feelings are a bit hurt by Jesus's need to repeatedly ask for his love) we can imagine the warm relief of reconciliation in the wake of it.

Rowan Williams observes that when Jesus speaks to Peter, he calls him "Simon, son of John," his name the first time that Jesus encountered him (John 1:42).[22] It is interesting, as Williams notes, that the disciples have gone back to their original circumstances, as if their period of discipleship and the trauma of the crucifixion had never happened.[23] But while there is again a biblical repetition here, a re-playing of that which has gone before, it is not as if the disciples can erase all that has occurred, the "past of their desertion and failure."[24] Peter, in particular, is "the failed apostle . . . [who] has to recognize himself as betrayer: that is part of the past that makes him who he is."[25] Williams writes, "If [Simon] is to be called again, if he can again become a true apostle, the 'Peter' that he is in the purpose of Jesus rather than the Simon who runs back into the cosy obscurity of 'ordinary' life, his failure must be assimilated, lived through again and brought to good and not to destructive issue."[26]

This human failure thus becomes utterly woven into the story of the resurrection. Indeed, in the lack of a heroic figure striding forth from the tomb, the story of the resurrection *is* this scene of forgiveness on the beach. And it is not only Peter who must move forward through his failure. Williams writes that Jesus "comes now to men whose history is one of initial hope and promise, followed by betrayal and emptiness. They are called now and sent now as forgiven men . . . . On the far side of the resurrection, vocation and forgiveness occur together, always and inseparably."[27] This is important:

> Peter's fellowship with the Lord is not over, not ruined, it still exists and is alive because Jesus invites him to explore it further. Here the past is returned within a lived relationship that is evidently moving and growing. To know that Jesus still invites is to know

22. Williams, *Resurrection*, 28.

23. Williams, *Resurrection*, 27–28.

24. Williams, *Resurrection*, 28.

25. Williams, *Resurrection*, 28.

26. Williams, *Resurrection*, 28–29.

27. Williams, *Resurrection*, 29.

that he accepts, forgives, bears and absorbs the hurt done: to hear the invitation is to know oneself forgiven, and *vice versa.*

Thus the memory of failure is in this context the indispensable basis of a calling forward in hope. Peter, in being present to Jesus, becomes—painfully and nakedly—present to himself: but that restoration to him of an identity of failure is also the restoration of an identity of hope. The presence of Jesus, still faithful, still calling, inviting his followers to love him, opens out the past in grace. And what Peter may learn is that wherever he may find himself, however he may fall, his life is constantly capable of being opened to God's creative grace: God's presence in Jesus will not fail him. The inconstant, vulnerable decisions and commitments of human beings, endlessly liable to destructive illusion, are set against a backcloth of God's constant decision and external commitment, his everlasting invitation to and 'making space' for his creatures.[28]

The story of the resurrection, then, is in many ways the story of human resurrections, of a pattern of fall and return, of failure and restoration. Seeing ourselves in Peter, we can recognize how our "failure must be assimilated, lived through again and brought to good and not to destructive issue."

When Peter realizes that it is Jesus who is on the shore, he is not filled with joy and hope but casts himself into the sea. We might almost picture his dive like the little kicking legs in "Landscape with Icarus." Where once he walked on water (Matt 14:29), Peter now sinks with a sense of failure. But at this point Jesus also is not walking on water or floating up to heaven on angels' wings, an inverse image of Icarus's fall. Jesus is quietly breaking bread on the beach.

The hope of the resurrection, as we strive to work for the planet and a healthy life for our inheritors, is thus not that everything will be resolved with a miraculous scene of overthrowing death and destruction, but that in spite of our personal and collective failures we move forward into a better life. Our manifold ecological sins do not mean that we have to give up, that we need to cast ourselves into the sea in despair, that we need to hide in a state of environmental shame. Our propensity to be "endlessly liable to destructive illusion"—such as the illusion that we can consume nature's resources without destroying nature itself—can be repented and forgiven. We need to change and be changed, but our failings are part of the process, not a source of determinate exclusion from grace.

28. Williams, *Resurrection*, 30.

And the work is not solitary. Jesus joins the fishermen in their labors, and everyone sits together as they have their peaceful seaside meal. We are tempted to picture the resurrection as a single event, as the individual person of Jesus springs from the tomb. But as we have seen, that is our own created illusion—none of the gospel narratives have described that to us. Instead we have the emptiness of the tomb and the emptiness of that part of the story, as the narrative drive is re-directed into scenes of communality, of Jesus walking and eating with friends. Amy Frykholm has discussed how the Pauline vision of the world that was created by the resurrection inverts an ancient warrior myth.[29] Frykholm notes that "in this myth, the people face a crisis and are saved by a man who is stronger and more powerful than they are, a man who is in some way divine. In Paul's inversion, the weakness of Christ—which is an unearthly power—saves."[30] The absence of a strong hero is part of Paul's larger vision of the kingdom of God established by the resurrection, a kingdom that is both already and not-yet. "The new age was imminent and required the absolute transformation of the believer. From the resurrection of Jesus, a new community had been born, and that community was the inversion of the imperial society in which it was located. Paul built an image of this community through a series of contrasts": God's power rests in a crucified criminal, not the Roman emperor; the kingdom of God is inherited by the weak, not the strong; wisdom lies in foolishness, not the worldly wisdom of the philosophers.[31]

If, in working towards a better world for our children and our inheritors, we feel weak or foolish, that is part of the resurrection narrative. The powers of empire are probably not going to save us, and we will not have a solitary climate hero. If we are imperfect, and have made mistakes and continue to make mistakes, that is part of the resurrection narrative—to return to Williams's powerful words, "the memory of failure is . . . the indispensable basis of a calling forward in hope." If we cannot walk on water and we periodically even lose our faith, that, too, is part of the resurrection narrative. This is not a story of one-time catastrophic failure—as in the case of Icarus, whose reach for the sun left him plummeting into the sea—but the steady call to work towards the kingdom of God on earth.

29. Frykholm, *Christian Understandings of the Future*, 58.

30. Frykholm, *Christian Understandings of the Future*, 58.

31. Frykholm, *Christian Understandings of the Future*, 58.

Epilogue
Getting on With It

But each of us was given grace according to the measure of Christ's gift. . . . The gifts he gave were that some would be apostles, some prophets, some evangelists, some pastors and teachers, to equip the saints for the work of ministry, for building up the body of Christ, until all of us come to the unity of the faith and of the knowledge of the Son of God, to maturity, to the measure of the full stature of Christ. We must no longer be children, tossed to and fro and blown about by every wind of doctrine, by people's trickery, by their craftiness in deceitful scheming. But speaking the truth in love, we must grow up in every way into him who is the head, into Christ, from whom the whole body, joined and knit together by every ligament with which it is equipped, as each part is working properly, promotes the body's growth in building itself up in love.

—Ephesians 4:7, 11–16

Do what you can, with what you've got, where you are.

—Squire Bill Widener[1]

The subtitle of this book ("to give a future with hope") comes from a passage in the book of Jeremiah. God is speaking to the Jewish people who are living in captivity in Babylon. God urges these Israelites to continue with their lives—to build houses, plant gardens, and raise children. Although they are exiles in a foreign land, God tells them to not only make peace

---

1. This quote is often falsely attributed to Theodore Roosevelt, who quoted it in his autobiography; see Brewton, "Squire Bill Widener." It is the best life advice ever.

with where they are, but to work for the prospering of the city: "seek the welfare of the city where I have sent you into exile, and pray to the LORD on its behalf, for in its welfare you will find your welfare" (Jer 29:7). God assures them that he continues to look out for them during this period of exile. "I know the plans I have for you, says the LORD, plans for your welfare and not for harm, to give you a future with hope" (Jer 29:11). God asks for their trust, but this is not a passive faith—rather, their faith in the future is manifest through the activities of their daily lives, as they make homes and harvest the fields, as they celebrate marriages and see people come into this world and leave it. Basically, he tells them to get on with it, as the British might say. A future with hope is thus something in which we put our faith, even as we work towards it. This hope-filled future requires deep trust in ultimate goodness and a commitment to strive for the welfare of our communities.

On my dark days, I lose hope. We have already squandered so much time that we might have spent getting ahead of this climate crisis; the environmental prognosis is not rosy, especially as the years slip by and we continue to increase our planetary emissions instead of decreasing them. There are powerful political forces that stand in the way of meaningful change, even when polls increasingly show that a majority of citizens are concerned about the consequences of a warming planet and are supportive of measures like pricing carbon pollution. I get tired. I get tempted by despair.

But as a person of faith, I try to remind myself of God's promise of a light-filled future, and of our responsibility to work towards this future. This pairing of covenant and obligation does not just appear in Jeremiah, but is woven throughout the Scripture and tradition. It is even at the heart of our shared prayer: "thy kingdom come, thy will be done" (Matt 6:10). So how do we get on with doing God's will?

I have found three biblical passages particularly helpful in shaping my own response to the climate crisis. First, there is the profound wisdom of First Corinthians, where Paul recognizes that there are a variety of talents and abilities (charisms) that can be put to use for the community, and he validates and welcomes them all:

> Now there are varieties of gifts, but the same Spirit; and there are varieties of services, but the same Lord; and there are varieties of activities, but it is the same God who activates all of them in everyone. To each is given the manifestation of the Spirit for the common good. To one is given through the Spirit the utterance of

wisdom, and to another the utterance of knowledge according to
the same Spirit, to another faith by the same Spirit, to another gifts
of healing by the one Spirit, to another the working of miracles, to
another prophecy, to another the discernment of spirits, to another
various kinds of tongues, to another the interpretation of tongues.
All these are activated by one and the same Spirit, who allots to
each one individually just as the Spirit chooses. For just as the body
is one and has many members, and all the members of the body,
though many, are one body, so it is with Christ. (1 Cor 12:4–12)

"To each is given the manifestation of the Spirit for the common good."
Everyone has a spiritual gift that, in turn, is to be given to the community.
There is so much work to do, and none of us can do it all. We are called
to discern our gift, and to use it appropriately. (This idea is similarly expressed in Paul's letter to the Ephesians, quoted in this chapter's epigraph.)
The use of our particular gifts is also important for the mission of reducing
our carbon emissions. Are you good at engineering, accounting, listening,
talking, protesting, farming, gardening, hosting, praying, homemaking,
governing, marketing, building, inventing, teaching, preaching, studying,
creating, leading, following, persuading, advocating, or supporting? (And
the list goes on . . .) There is a role in this for everyone. I figure I am good
at thinking and writing. Here is my book, my small contribution towards
the common good.

Secondly, there is the rather quirky parable that Jesus tells the disciples
in Luke 18. In this story, there is a judge who has no fear of God and no
respect for other people. In his town there is a widow (a person on the
margins with no political or social power) who keeps demanding justice
for her cause. Eventually, the judge relents to the widow's pleas—not from
admirable motives but simply to make her go away, "so that she may not
wear [him] out by continually coming" (Luke 18:5). The parable is a vali-
dation of, well, righteous nagging. Jesus offers this parable as a reminder
to his disciples of "their need to pray always and not to lose heart" (Luke
18:1). I can sometimes lose heart: I am tired, and busy, and I don't always
understand why others don't see the need for urgent climate action. I start
to feel that my efforts are pointless. But the moral of this odd biblical story,
essentially, is to keep bothering people (and God) in the push for justice.

I had to chuckle when this parable was the Gospel reading on the
Sunday after my church had finally agreed to switch our electricity contract
from fossil fuels to 100 percent renewables. It took nine months for everyone to agree to the change, as the discussion bounced between a climate

change reading group, the Property Committee, the Assistant Rector, the Finance Committee, the Rector, the church secretary, the utility vendor, and the Vestry (i.e., the lay leadership board). My church does not have a godless judge at its helm; it is full of good, well-intentioned people. Yet in discussions about the move to renewable energy I found myself needing to play the widow of the parable—asking, reminding, petitioning, meeting, nudging, e-mailing, advocating, prodding, and finally thanking. On more than one occasion I was tempted to give up. In the end, in spite of my impatience and irritation, I learned a valuable lesson about group dynamics: the involvement of the whole community was the wise approach (thanks for the guidance, Albert and Joseph). Now everyone can flip on a light switch in the church kitchen, Sunday school classroom, or sanctuary and know that for our 125,000 kilowatt hours per year we are doing our part to reduce carbon emissions. (In what feels like either a divine test or a divine joke, a decimal error in figures provided by our electricity vendor led us to understand that the shift to renewables would cost the church $1,200 more per year; after we decided to do the right thing in spite of the financial sacrifice, we learned that the shift would only be $120 more per year, a miniscule sum in the context of the operating budget. The cost of renewable energy has plummeted in recent years.)

Much of the climate work that needs to be done in our communities—our churches/synagogues/mosques, our campuses, our office buildings, our towns, our states, our nations, and our planet—is unglamorous and requires a steady chorus of civic nagging. As someone who just likes to get something done and move on to the next big thing, I have been trying to learn the merits of patience—with myself, with others, and with process. I have been deeply inspired by the persistence, resilience, commitment, and stamina of veterans of the environmental movement.[2]

2. Just to mention some people who inspire me through my work with the Citizens' Climate Lobby: in the Philadelphia chapter, Peter Handler, Karen Melton, and Sarah Davidson, who keep the wheels steadily moving forward; in the national Episcopal Action Team, Emily Hopkins, who led the national Episcopal Church to pass a resolution endorsing a carbon fee and dividend model of carbon pricing; in the national CCL organization, the talented Mark Reynolds and Danny Richter, who clearly could have chosen lucrative careers and days on the golf course, but have instead committed themselves to fight for a livable world; and the staff of the national CCL organization, who have impressed me with their dedication and kindness. As I was finalizing this book for publication, I learned of the death of CCL's founder, Marshall Saunders; I have been inspired by his example of vision, initiative, and generosity of spirit.

And finally, when I am feeling low on hope and hard of heart, I find myself returning to a third Bible verse, a favorite of my late beloved dissertation director: "He has told you, O mortal, what is good; / and what does the LORD require of you / but to do justice, and to love kindness, / and to walk humbly with your God?" (Mic 6:8). As I have learned more and more about climate change, I have often found myself becoming judgmental of others' choices. For instance, why does nearly everyone in the car line at my daughter's school drive a large SUV? Some people perhaps have not thought about the environmental impact of their choice of vehicle; some people may be deliberately and ostentatiously flouting climate science; but many people just drive the carpool or the volleyball team, and need to seat many children. I try to imagine my way into other people's circumstances and to suspend my rush to judgment. Walking humbly also requires humility in seeing the specks, or even the logs or beams, in our own eye (Matt 7:3). Why do I choose to live so far from my place of work, requiring a long commute? Why do I fly to so many conferences? Why, now that my home electricity is 100 percent renewable, do I cling to my kitchen stove which is fueled by natural gas? People are living their lives, with their unique set of circumstances and stressors, usually as best as they can figure to do. My hope is that, as a people, we will notice our carbon conditions more, and that environmentally helpful choices will also become more and more naturally integrated into our lives. (This is the economic logic of carbon pricing.) For now I have to remind myself that exhausted parents and stressed out breadwinners and people who are grieving, anxious, depressed, or otherwise downtrodden are often just trying to get through the day, much less consider the existential consequences of global climate change.

Throughout this book I have cited many statistics, but for the mission of this project the most critical figure is that 9 percent: the very low percentage of Americans who consider climate change a religious issue.[3] I hope that as more and more people become aware of the human toll of climate change (that the issue isn't just about protecting animals and nature, although those are crucial, too), more people of faith will connect the dots to religious teachings that call for care of the neighbor. While I have been writing from my own Christian point of view, Christianity is a very large tent. Here are some organizations that might match your particular religious identity:

3. Yale Program on Climate Change Communications, "Climate Change in the American Mind."

- Global Catholic Climate Movement
- Evangelical Environmental Network
- Interfaith Power and Light

Christianity is fundamentally a communal religion, founded, in a way, around a dinner table. We can all do our part as individuals, but we have a louder voice in communion with each other, with other congregations, with other faith traditions. Addressing climate change is not just a political or an economic issue; it is a moral and spiritual one. People of faith should raise their voices and join together. The Lord requires us to do justice.

Bibliography

Accordance Bible Study Software, version 12.3.2. Altamonte Springs, FL: Oak Tree Software Inc., 1999.

Akerlof, George, et al. "Economists' Statement on Carbon Dividends." *The Wall Street Journal*, January 16, 2019. https://www.wsj.com/articles/economists-statement-on-carbon-dividends-11547682910.

Albeck-Ripka, Livia. "Koala Mittens and Baby Bottles: Saving Australia's Animals After Fires." *The New York Times*, January 7, 2020. https://www.nytimes.com/2020/01/07/world/australia/animals-wildlife-fires.html?action=click&module=News&pgtype=Homepage.

American Chemical Society. "A Greenhouse Effect Analogy." Accessed October 5, 2018. https://www.acs.org/content/acs/en/climatescience/climatesciencenarratives/a-greenhouse-effect-analogy.html.

Antal, Jim. *Climate Church, Climate World: How People of Faith Must Work for Change.* Lanham, MD: Rowman and Littlefield, 2018.

Auden, W. H. *Collected Poems*, edited by Edward Mendelson. New York: Vintage International, 1991.

Augustine. *On the Trinity, Books 8–15*, edited by Gareth Matthews. Translated by Stephen McKenna. Cambridge Texts in the History of Philosophy. Cambridge: Cambridge University Press, 2002.

Australian Government, Department of Environment and Energy. "Greenhouse Effect." Accessed December 9, 2019. http://www.environment.gov.au/climate-change/climate-science-data/climate-science/greenhouse-effect.

Avril, Tom. "Climate Change is Hurting Philadelphians' Health, and the Worst is Yet to Come." *Philadelphia Inquirer*, September 11, 2019. https://www.inquirer.com/science/climate/climate-change-heat-death-heart-failure-20190911.html?utm_medium=email&utm_campaign=Morning%20email%2009-12-19P&utm_content=Morning%20email%2009-12-19P+CID_96d4c036b00240d7fac6ace3fea9c846&utm_source=email&utm_term=Its%20a%20question%20of%20when.

Backus, Irena. *Reformation Readings of the Apocalypse: Geneva, Zurich, and Wittenberg.* Oxford Studies in Historical Theology. New York: Oxford University Press, 2000.

Banis, Davide. "10 Worst Climate-Driven Disasters of 2018 Cost $85 Billion." *Forbes*, December 28, 2018. https://www.forbes.com/sites/davidebanis/2018/12/28/10-worst -climate-driven-disasters-of-2018-cost-us-85-billion/#1fefa2422680.

Barnard, Anne, and Josh Haner. "Climate Change is Killing the Cedars of Lebanon." *New York Times*, July 18, 2018. https://www.nytimes.com/interactive/2018/07/18/ climate/lebanon-climate-change-environment-cedars.html.

Barron, Jesse. "How Big Business is Hedging Against the Apocalypse." *New York Times*, April 11, 2019. https://www.nytimes.com/interactive/2019/04/11/magazine/climate -change-exxon-renewable-energy.html.

Bell, Mark R. *Apocalypse How? Baptist Movements During the English Revolution.* Macon, GA: Mercer University Press, 2000.

Bloomberg, Michael, and Carl Pope. *Climate of Hope: How Cities, Businesses, and Citizens Can Save the Planet.* New York: St. Martin's, 2017.

Bon Appétit Management Company. "Tackling Climate Change Through Our Food Choices." Accessed October 5, 2018. http://www.bamco.com/timeline/low-carbon- diet/.

Boren, Michael. "Philadelphia Is Getting Hotter, Wetter, and Snowier at the Same Time." *Philadelphia Inquirer*, August 2, 2018. http://www2.philly.com/philly/news/climate- change-philadelphia-weather-temperatures-rainfall-snowfall-20180802.html.

Borsch, Frederick. *The Spirit Searches Everything: Keeping Life's Questions.* Cambridge, MA: Cowley, 2005.

Bouma-Prediger, Steven. *For the Beauty of the Earth: A Christian Vision for Creation Care.* 2nd ed. Grand Rapids, MI: Baker Academic, 2010.

Bradshaw, Paul F., and Maxwell E. Johnson. *The Eucharistic Liturgies: Their Evolution and Interpretation.* Collegeville, MN: Liturgical, 2012.

Brewton, Sue. "Squire Bill Widener vs. Theodore Roosevelt." *Sue Brewton's Blog: On Quotes and Misquotes*, December 31, 2014. https://suebrewton.com/tag/do-what- you-can-with-what-you-have-where-you-are/.

Burke, Minyvonne. "Video Shows Koalas, Other Animals Hurt in Australia's Fires Getting Treated." NBC News, January 10, 2020. https://www.nbcnews.com/news/world/ video-shows-koalas-other-animals-hurt-australia-s-fires-getting-n1113436.

Carson, Rachel. *Silent Spring.* Boston: Mariner, 2002; 1962.

Celsus. *On the True Doctrine: A Discourse Against the Christians.* Translated by R. Joseph Hoffmann. Oxford: Oxford University Press, 1987.

Chester, Tim. *Mission and the Coming of God: Eschatology, the Trinity and Mission in the Theology of Jürgen Moltmann and Contemporary Evangelism.* Paternoster Theological Monographs. Eugene, OR: Wipf & Stock, 2006.

Clark, Duncan. "How Long Do Greenhouse Gases Stay in the Air?" *The Guardian*, January 16, 2012. https://www.theguardian.com/environment/2012/jan/16/greenhouse-gases -remain-air.

Collomb, Jean-Daniel. "The Ideology of Climate Denial in the United States." *European Journal of American Studies* 9.1 (2014) 1–20. https://journals.openedition.org/ ejas/10305.

Congdon, David W. *The God Who Saves: A Dogmatic Sketch.* Kindle edition. Eugene, OR: Cascade, 2016.

Court, John M. *Approaching the Apocalypse: A Short History of Christian Millenarianism.* New York: I. B. Tauris, 2008.

Dalferth, Ingolf U. *Umsonst.* Tübingen: Mohr Siebeck, 2011.

Dalmais, Irénée Henri. "Introduction." In *The Church at Prayer: An Introduction to the Liturgy,* Vol. 4, *The Liturgy and Time,* by Irénée Henri Dalmais, Pierre Jounel and Aimé Georges Martimort, 1–7. Translated by Matthew J. O'Connell. Collegeville, MN: Liturgical, 1986.

De Coninck, H., and A. Revi, M. Babiker, P. Bertoldi, M. Buckeridge, A. Cartwright, W. Dong, J. Ford, S. Fuss, J.-C. Hourcade, D. Ley, R. Mechler, P. Newman, A. Revokatova, S. Schultz, L. Steg, T. Sugiyama. "Strengthening and Implementing the Global Response." In *Global Warming of 1.5°C. An IPCC Special Report on the impacts of global warming of 1.5°C above pre-industrial levels and related global greenhouse gas emission pathways, in the context of strengthening the global response to the threat of climate change, sustainable development, and efforts to eradicate poverty* (2018), edited by V. Masson-Delmotte, P. Zhai, H.-O. Pörtner, D. Roberts, J. Skea, P.R. Shukla, A. Pirani, W. Moufouma-Okia, C. Péan, R. Pidcock, S. Connors, J.B.R. Matthews, Y. Chen, X. Zhou, M.I. Gomis, E. Lonnoy, T. Maycock, M. Tignor, and T. Waterfield. https://www.ipcc.ch/site/assets/uploads/sites/2/2019/02/SR15_Chapter4_Low_Res.pdf.

De-Shalit, Avner. *Why Posterity Matters: Environmental Policy and Future Generations.* Environmental Philosophies Series. London: Routledge, 1995.

Deneen, Patrick J. *Why Liberalism Failed.* Kindle edition. New Haven: Yale University Press, 2018.

Desjardins, Joseph R. *Environmental Ethics: An Introduction to Environmental Philosophy.* 5th ed. Boston: Wadsworth, Cengage Learning, 2013.

Di Cesare, Mario A., ed. *George Herbert and the Seventeenth-Century Religious Poets: Authoritative Texts and Criticism.* New York: W. W. Norton and Co., 1978.

Dobson, Andrew, ed. *Fairness and Futurity: Essays on Environmental Sustainability and Social Justice.* Oxford: Oxford University Press, 1999.

Donne, John. *Devotions Upon Emergent Occasions.* In *John Donne,* edited by Janel Mueller. 21st-Century Oxford Authors. Oxford: Oxford University Press, 2015.

Druke, Galen. "End of Monarch Butterfly Migration Could Be in Sight." Wisconsin Public Radio, January 30, 2014. https://www.wpr.org/end-monarch-butterfly-migration-could-be-sight.

Editorial Board. "Is It Time to Cancel the Climate-Change Apocalypse?" *Richmond Times-Dispatch,* June 18, 2015. https://www.richmond.com/opinion/our-opinion/editorial-is-it-time-to-cancel-the-climate-change-apocalypse/article_8b5191ec-12cb-5289-9946-4a6448519d8b.html.

Editorial Board. "Not the Climate Apocalypse." *The Wall Street Journal,* August 21, 2018. https://www.wsj.com/articles/not-the-climate-apocalypse-1534894336.

EPA [United States Environmental Protection Agency]. "Sources of Greenhouse Gas Emissions." Accessed January 12, 2020. https://www.epa.gov/ghgemissions/sources-greenhouse-gas-emissions.

Episcopal Church. The Book of Common Prayer. New York: Oxford University Press, 1990.

"Flood Waters Unearth Caskets in Michigan Cemetery." WTHR-NBC, June 11, 2018. http://www.wthr.com/article/flood-waters-unearth-caskets-in-michigan-cemetery.

Francis (Pope). *Laudato Si': On Care for Our Common Home.* Huntington, IN: Our Sunday Visitor, 2015.

Freedman, Andrew. "Extreme Weather Patterns Are Raising the Risk of a Global Food Crisis, and Climate Change Will Make This Worse." *The Washington Post,* December

9, 2019. https://www.washingtonpost.com/weather/2019/12/09/extreme-weather-patterns-are-raising-risk-global-food-crisis-climate-change-will-make-this-worse/.

Friedrich, Otto. *The End of the World: A History*. New York: Coward, McCann & Geoghegan, 1982.

Frykholm, Amy. *Christian Understandings of the Future: The Historical Trajectory*. Kindle Edition. Minneapolis: Fortress, 2016.

Geagan, Kate. *Go Green Get Lean: Trim Your Waistline with the Ultimate Low-Carbon Footprint Diet*. New York: Rodale, 2009.

Gearino, Dan. "U.S. Wind Power Is 'Going All Out' with Bigger Tech, Falling Prices, Reports Show." *Inside Climate News*, August 23, 2018. https://insideclimatenews.org/news/23082018/wind-energy-prices-market-growth-offshore-tax-credits-turbines-technology.

Geary, Patrick J. *Living with the Dead in the Middle Ages*. Ithaca: Cornell University Press, 1994.

Gershon, David. *Low Carbon Diet: A 30 Day Program to Lose 5000 Pounds*. Third revised edition. Woodstock, NY: Empowerment Institute, 2007.

Golley, Frank Benjamin. *A History of the Ecosystem Concept in Ecology: More Than the Sum of the Parts*. New Haven: Yale University Press, 1993.

Greenblatt, Stephen, et al., eds. *The Norton Shakespeare, Based on the Oxford Edition*. New York: W. W. Norton & Co., 1997.

Guskin, Emily, et al. "Americans Broadly Accept Climate Science, But Many Are Fuzzy on the Details." *The Washington Post*, December 9, 2019. https://www.washingtonpost.com/science/americans-broadly-accept-climate-science-but-many-are-fuzzy-on-the-details/2019/12/08/465a9d5e-0d6a-11ea-8397-a955cd542d00_story.html.

Gustin, Georgina. "Investors Join Calls for a Food Revolution to Fight Climate Change." *Inside Climate News*, January 29, 2019. https://insideclimatenews.org/news/29012019/global-food-system-shocks-climate-change-mcdonalds-obesity-malnutrition-investors-lancet-scientists.

———. "U.S. Taxpayers on the Hook for Insuring Farmers Against Growing Climate Risks." *Inside Climate News*, December 31, 2018. https://insideclimatenews.org/news/31122018/crop-insurance-farm-bureau-taxpayer-subsidies-climate-change-risk-rising?utm_source=InsideClimate+News&utm_campaign=bfded7af77-&utm_medium=email&utm_term=0_29c928ffb5-bfded7af77-327866781.

Hannibal, Mary Ellen. "Are We Watching the End of the Monarch Butterfly?" *New York Times*, January 25, 2019. https://www.nytimes.com/2019/01/25/opinion/monarch-butterfly-california-extinction.html.

Hansen, James. *Storms of My Grandchildren: The Truth About the Coming Climate Catastrophe and Our Last Chance to Save Humanity*. New York: Bloomsbury, 2009.

Harper, Tim. "Canadians End Up the Big Losers in the Apocalyptic Climate Change Debate." *The Star*, May 8, 2018. https://www.thestar.com/opinion/star-columnists/2018/05/08/canadians-end-up-the-big-losers-in-the-apocalyptic-climate-change-debate.html.

Harrill, J. A. "Stoic Physics, the Universal Conflagration, and the Eschatological Destruction of the 'Ignorant and Unstable' in 2 Peter." In *Stoicism in Early Christianity*, edited by Tuomas Rasimus, Troels Engberg-Pedersen, and Ismo Dunderberg, 115–140. Peabody, MA: Henrickson, 2010.

Hayes, Christopher M., in collaboration with Brandon Gallaher, Julia S. Konstantinovsky, Richard J. Ounsworth, and C. A. Strine. *When the Son of Man Didn't Come: A Constructive Proposal on the Delay of the Parousia.* Minneapolis: Fortress, 2016.

Hayhoe, Katharine, and Andrew Farley. *A Climate for Change: Global Warming Facts for Faith-Based Decisions.* New York: FaithWords, 2009.

Hescox, Mitch, and Paul Douglas. *Caring for Creation: The Evangelical's Guide to Climate Change and a Healthy Environment.* Bloomington, MN: Bethany House, 2016.

Holinshed, Raphael, et al. "Richard the Second, the Second Sonne to Edward Prince of Wales." In *The Chronicles of England, Scotland and Ireland.* 6 vols. London, 1587. http://english.nsms.ox.ac.uk/holinshed/texts.php?text1=1587_4380.

Isaac, Rael Jean. *Roosters of the Apocalypse: How the Junk Science of Global Warming is Bankrupting the Western World.* 2nd ed. CreateSpace Independent Publishing Platform, 2013.

Jarvis, Brooke. "The Insect Apocalypse is Here." *New York Times,* November 27, 2018. https://www.nytimes.com/2018/11/27/magazine/insect-apocalypse.html.

Jennings, Nathan G. *Liturgy and Theology: Economy and Reality.* Eugene, OR: Cascade, 2017.

Johnston, Warren. *Revelation Restored: The Apocalypse in Later Seventeenth-Century England.* Studies in British Religious History 27. Woodbridge, UK: Boydell, 2011.

Jonas, Hans. *The Imperative of Responsibility: In Search of an Ethics for the Technological Age.* Translated by Hans Jonas with David Herr. Chicago: The University of Chicago Press, 1984.

Jounel, Pierre. "The Easter Cycle." In *The Church at Prayer: An Introduction to the Liturgy* Vol. 4, *The Liturgy and Time,* by Irénée Henri Dalmais, Pierre Jounel and Aimé Georges Martimort, 33–76. Translated by Matthew J. O'Connell. Collegeville, MN: Liturgical, 1986.

Käsemann, Ernst. *Essays on New Testament Themes.* Studies in Biblical Theology 41. London: SCM, 1964.

Kavka, Gregory. "The Futurity Problem." In *Obligations to Future Generations,* edited by R. I. Sikora and Brian Barry, 186–203. Philosophical Monographs, Second Series. Philadelphia: Temple University Press, 1978.

Koren, Marina, and Robinson Meyer. "It's Colder Than Mars Out There." *The Atlantic,* December 29, 2017. https://www.theatlantic.com/science/archive/2017/12/cold-weather-united-states-mars/549386/.

Krahenbuhl, Peter Davis. "Last Butterfly Migration?" *Good Nature Travel, Natural Habitat Adventures Blog of the World Wildlife Foundation,* July 5, 2017. http://goodnature.nathab.com/last-monarch-butterfly-migration/.

Landes, Richard. "The Fruitful Error: Reconsidering Millennial Enthusiasm." Review of *Apocalypses: Prophecies, Cults, and Millennial Beliefs through the Ages,* Eugen Weber. *Journal of Interdisciplinary History* 32.1 (2001) 89–98.

Lapidge, Michael. "Stoic Cosmology." In *The Stoics,* edited by John M. Rist, 161–85. Berkeley: University of California Press, 1978.

Larsen, Matthew D. C. *Gospels Before the Book.* Oxford: Oxford University Press, 2018.

Lavelle, Marianne. "Americans Increasingly Say Climate Change is Happening Now." *Inside Climate News,* January 23, 2019. https://insideclimatenews.org/news/22012019/climate-change-survey-impact-now-americans-extreme-weather-george-mason-yale.

Laville, Sandra, and Matthew Taylor. "A Million Bottles a Minute: World's Plastic Binge 'As Dangerous as Climate Change.'" *The Guardian*, June 28, 2017. https://www.theguardian.com/environment/2017/jun/28/a-million-a-minute-worlds-plastic-bottle-binge-as-dangerous-as-climate-change.

Lewis, C. S. "The World's Last Night." In *The World's Last Night and Other Essays*, by C. S. Lewis, 93–113. New York: Harcourt, Brace and Company, 1952.

Linden, Eugene. "How Scientists Got Climate Change So Wrong." *New York Times*, November 8, 2019. https://www.nytimes.com/2019/11/08/opinion/sunday/science-climate-change.html.

McKibben, Bill. "The Question I Get Asked the Most." *EcoWatch*, October 14, 2016. https://www.ecowatch.com/bill-mckibben-climate-change-2041759425.html.

———. "This is How Human Extinction Could Play Out." *Rolling Stone*, April 9, 2019. https://www.rollingstone.com/politics/politics-features/bill-mckibben-falter-climate-change-817310/.

Miklósházy, Attila. *The Origin and Development of the Christian Liturgy According to Cultural Epochs: Political, Cultural and Ecclesial Backgrounds: History of the Liturgy.* Vol. 1. Lewiston, NY: Edwin Mellen, 2006.

Milton, John. *John Milton: Complete Poems and Major Prose*, edited by Merritt Y. Hughes. Indianapolis: Odyssey, 1957.

Moe-Lobeda, Cynthia. "Climate Injustice, Truth-telling, and Hope." *Anglican Theological Review* 99.3 (2017) 531–40.

Moltmann, Jürgen. *History and the Triune God: Contributions to Trinitarian Theology.* Translated by John Bowden. New York: Crossroad, 1992.

———. *The Trinity and the Kingdom: The Doctrine of God.* Translated by Margaret Kohl. San Francisco: Harper & Row, 1981.

Moore, Arthur Lewis. *The Parousia in the New Testament.* Leiden: E. J. Brill, 1966.

Mora, Camilo, Daniele Spirandelli, Erik C. Franklin, John Lynham, Michael B. Kantar, Wendy Miles, Charlotte Z. Smith, Kelle Freel, Jade Moy, Leo V. Louis, Evan W. Barba, Keith Bettinger, Abby G. Frazier, John F. Colburn IX, Naota Hanasaki, Ed Hawkins, Yukiko Hirabayashi, Wolfgang Knorr, Christopher M. Little, Kerry Emanuel, Justin Sheffield, Jonathan A. Patz, and Cynthia L. Hunter. "Broad Threat to Humanity from Cumulative Climate Hazards Intensified by Greenhouse Gas Emissions." *Nature Climate Change* 8.12 (2018) 1062–71.

Mouhot, Jean-François. "In Pursuit of the Apocalypse." *History Today* 62.8 (2012) 6–7.

Nace, Trevor. "We're Now at a Million Plastic Bottles Per Minute—91% Of Which Are Not Recycled." *Forbes*, July 26, 2017. https://www.forbes.com/sites/trevornace/2017/07/26/million-plastic-bottles-minute-91-not-recycled/#5e8e57b2292c.

NASA [National Aeronautics and Space Administration]. "2019 Ozone Hole is the Smallest on Record Since Its Discovery." Accessed December 12, 2019. https://www.nasa.gov/feature/goddard/2019/2019-ozone-hole-is-the-smallest-on-record-since-its-discovery.

———. "Scientific Consensus: Earth's Climate is Warming." Accessed November 17, 2019. https://climate.nasa.gov/scientific-consensus/.

National Resources Defense Council. "Health and Climate Change: Accounting for Costs." November 2011. https://www.nrdc.org/sites/default/files/accountingcosts.pdf.

National Wildlife Federation. "Monarchs Face New Threats, Losses Along Migration Route." *National Wildlife Federation Blog*, January 31, 2014. https://blog.nwf.org/2014/01/monarchs-face-new-threats-losses-along-migration-route/.

New Revised Standard Version (NRSV) Bible. Division of Christian Education and the National Council of the Churches in Christ in the United States of America, 1989. Used with permission in Accordance Bible software, version 12.3.2.

Northcott, Michael S. *The Environment and Christian Ethics*. New Studies in Christian Ethics. Cambridge: Cambridge University Press, 1996.

———. *A Moral Climate: The Ethics of Global Warming*. London: Darton, Longman and Todd; Maryknoll, NY: Orbis, 2007.

Olson, Roger E. "Embarrassed by the Parousia? Evangelicals and the Return of Christ." *Patheos*, September 20, 2015. https://www.patheos.com/blogs/rogereolson/2015/09/embarrassed-by-the-parousia-evangelicals-and-the-return-of-christ/#disqus_thread.

Oxford English Dictionary Online. Oxford University Press, 2019. http://www.oed.com/.

Palmer, James T. *The Apocalypse in the Early Middle Ages*. New York: Cambridge University Press, 2014.

Pierre-Louis, Kendra. "Brace for the Polar Vortex: It May Be Visiting More Often." *New York Times*, January 28, 2019. https://www.nytimes.com/2019/01/18/climate/polar-vortex-2019.html.

———. "Why Is the Cold Weather So Extreme if the Earth Is Warming?" *New York Times*, January 31, 2019. https://www.nytimes.com/interactive/2019/climate/winter-cold-weather.html.

Poole, Kristen. "Always, Already, Again: Towards a New Typological Historiography." In *Early Modern Histories of Time: The Periodizations of Sixteenth- and Seventeenth-Century England*, edited by Kristen Poole and Owen Williams, 267–82. Philadelphia: University of Pennsylvania Press, 2019.

Popovich, Nadja. "How Does Your State Make Electricity?" *New York Times*, December 24, 2018. https://www.nytimes.com/interactive/2018/12/24/climate/how-electricity-generation-changed-in-your-state.html.

Renkl, Margaret. "The Last Butterfly." *New York Times*, September 3, 2018. https://www.nytimes.com/2018/09/03/opinion/to-save-monarch-butterfly-plant-milkweed-now.html.

Riebeek, Holli. "The Carbon Cycle." NASA Earth Observatory, June 16, 2011. https://earthobservatory.nasa.gov/features/CarbonCycle

Roberts, Karen B. "Driving Clean Energy Forward." *UDaily*, June 26, 2019. https://www.udel.edu/udaily/2019/june/vehicle-to-grid-integration-becomes-law/.

Robinson, Mary. *Climate Justice: Hope, Resilience, and the Fight for a Sustainable Future*. New York: Bloomsbury, 2018.

Rosenberg, Daniel, and Anthony Grafton. *Cartographies of Time: A History of the Timeline*. New York: Princeton Architectural, 2010.

Rousseau, Rob. "We Can't Stop the Climate Change Apocalypse Through Individual Action Alone." *Paste*, October 18, 2018. https://www.pastemagazine.com/articles/2018/10/we-cant-stop-the-climate-change-apocalypse-through.html.

Rowthorn, Jeffrey W., with W. Alfred Tisdale, Jr. *The Wideness of God's Mercy: Litanies to Enlarge Our Prayer*. Rev. ed. New York: Church, 2007.

Rumsey, Patricia M. *Sacred Time in Early Christian Ireland*. London: T & T Clark, 2007.

Sambursky, Samuel. *Physics of the Stoics*. London: Routledge and Paul, 1959.

Sawtell, Peter. "Stop Being an Individual." *Eco-Justice Notes*, May 20, 2016. http://www.eco-justice.org/E-160520.asp.

Scherer, Glenn. "Christian-right Views Are Swaying Politicians and Threatening the Environment." *Grist*, October 28, 2004. https://grist.org/article/scherer-christian/.

Schwartz, John. "Young People Are Suing the Trump Administration Over Climate Change. She's Their Lawyer." *New York Times*, October 23, 2018. https://www.nytimes.com/2018/10/23/climate/kids-climate-lawsuit-lawyer.html.

Sengupta, Somini. "Hotter, Drier, Hungrier: How Global Warming Punishes the World's Poorest." *New York Times*, March 12, 2018. https://www.nytimes.com/2018/03/12/climate/kenya-drought.html.

———. "Why Build Kenya's First Coal Plant? Hint: Think China." *New York Times*, February 27, 2018. https://www.nytimes.com/2018/02/27/climate/coal-kenya-china -power.html?action=click&contentCollection=Climate&module=RelatedCoverage &region=EndOfArticle&pgtype=article.

Senn, Frank C. *The People's Work: A Social History of the Liturgy*. Minneapolis: Fortress, 2010.

Serres, Michel, and Bruno Latour. *Conversations on Science, Culture, and Time*. Translated by Roxanne Lapidus. Ann Arbor: University of Michigan Press, 1995.

Severson, Kim. "From Apples to Popcorn, Climate Change is Altering the Foods America Grows," *New York Times*, April 30, 2019. https://www.nytimes.com/2019/04/30/ dining/farming-climate-change.html?smid=nytcore-ios-share.

Shore, Daniel. "Milton's Lonely God." *Milton Studies* 60.1–2 (2018) 29–52.

Sikora, R. I., and Brian Barry. "Introduction." In *Obligations to Future Generations*, edited by R. I. Sikora and Brian Barry, vii–xi. Philosophical Monographs, Second Series. Philadelphia: Temple University Press, 1978.

Sirna, Tony. "How You Can Go on a Carbon Diet." *Dancing Rabbit Ecovillage*. Accessed October 5, 2018. https://www.dancingrabbit.org/carbon-diet/.

Sleeth, J. Matthew. *Serve God, Save the Planet: A Christian Call to Action*. Grand Rapids, MI: Zondervan, 2006.

Stager, Lawrence E. "Jerusalem and the Garden of Eden." *Eretz-Israel: Archaeological, Historical and Geographical Studies* (1999) 183–94.

Staley, Samantha. "The Link Between Plastic Use and Climate Change: Essential Answer." *Stanford Magazine*, Nov/Dec 2009. https://stanfordmag.org/contents/the-link-between-plastic-use-and-climate-change-essential-answer.

Stoknes, Per Espen. *What We Think About When We Try Not to Think About Global Warming*. White River Junction, VT: Chelsea Green, 2015.

Subramanian, Meera. "Seeing God's Hand in the Deadly Floods, Yet Wondering about Climate Change." *Inside Climate News*, October 24, 2017. https://insideclimatenews. org/news/19102017/christianity-evangelical-climate-change-flooding-west-virginia.

Suttie, Jill. "How to Overcome 'Apocalypse Fatigue' Around Climate Change." *Greater Good*, February 23, 2018. https://greatergood.berkeley.edu/article/item/how_to_overcome _apocalypse_fatigue_around_climate_change.

Swyngedouw, Erik. "Apocalypse Forever? Post-Political Populism and the Spectre of Climate Change." *Theory, Culture and Society* 27.2–3 (2010) 213–32.

Tanner, Kathryn. *Economy of Grace*. Minneapolis, MN: Fortress, 2005.

Taylor, Charles. *A Secular Age*. Cambridge, MA: The Belknap Press of Harvard University Press, 2007.

Thanki, Nathan. "Fuck Your Apocalypse: Between Denial and Despair, a Better Climate Change Story." *The World at 1oC*, July 11, 2017. https://worldat1c.org/fuck-your-apocalypse-c82696b533d9.

Thoreau, Henry David. *Walden; or, Life in the Woods*. Boston: Ticknor and Field, 1854. Project Gutenberg e-book edition. https://www.gutenberg.org/files/205/205-h/205-h.htm.

———. *Wild Apples*. (1862). Project Gutenberg e-book edition. https://www.gutenberg.org/files/4066/4066-h/4066-h.htm.

Turkewitz, Julie, and Clifford Krauss. "In Colorado, a Bitter Battle Over Oil, Gas and the Environment Comes to a Head." *New York Times*, October 23, 2018. https://www.nytimes.com/2018/10/23/us/colorado-fracking-proposition-112.html.

University of Sydney, Australia. "A Statement About the 480 Million Animals Killed in NSW Bushfires Since September." January 3, 2020. https://sydney.edu.au/news-opinion/news/2020/01/03/a-statement-about-the-480-million-animals-killed-in-nsw-bushfire.html.

University of Rochester Newscenter. "Alien Apocalypse: Can Any Civilization Make It Through Climate Change?" June 4, 2018. https://www.rochester.edu/newscenter/astrobiology-alien-apocalypse-can-any-civilization-make-it-through-climate-change-322232/.

U.S. Global Change Research Program. *The Climate Report: The National Climate Assessment—Impacts, Risks, and Adaptation in the United States*. Brooklyn: Melville House, 2018.

Wade, Lizzie. "More Monarch Butterflies in Mexico, but Numbers Still Low." *Science*, January 27, 2015. https://www.sciencemag.org/news/2015/01/more-monarch-butterflies-mexico-numbers-still-low.

Wallace, David Foster. "This Is Water." YouTube, May 19, 2013. https://www.youtube.com/watch?reload=9&v=8CrOL-ydFMI.

Wang, J. M., C.-H. Jeong, N. Zimmerman, R. M. Healy, D. K. Wang, F. Ke, and G. J. Evans. "Plume-based Analysis of Vehicle Fleet Air Pollutant Emissions and the Contribution from High Emitters." *Atmospheric Measurement Techniques* 8 (2015) 3263–75. https://www.atmos-meas-tech.net/8/3263/2015/amt-8-3263-2015.pdf.

Weber, Eugen. *Apocalypses: Prophecies, Cults, and Millennial Beliefs through the Ages*. Cambridge, MA: Harvard University Press, 1999.

Whalen, Brett Edward. *Dominion of God: Christendom and Apocalypse in the Middle Ages*. Cambridge, MA: Harvard University Press, 2009.

Wikipedia. "Camp Fire," accessed January 22, 2018. https://en.wikipedia.org/wiki/Camp_Fire_(2018).

———. "Climate of Mars," accessed January 24, 2019. https://en.wikipedia.org/wiki/Climate_of_Mars.

———. "Landscape with the Fall of Icarus," accessed December 11, 2019. https://en.wikipedia.org/wiki/Landscape_with_the_Fall_of_Icarus

———. "List of Nonlinear Narrative Films," accessed December 9, 2019. https://en.wikipedia.org/wiki/List_of_nonlinear_narrative_films.

———. "Svante Arrhenius," accessed January 6, 2020. https://en.wikipedia.org/wiki/Svante_Arrhenius

Williams, Casey. "Apocalypse How? What Novels Screw Up About Climate Change." *Huffington Post*, April 21, 2018. https://www.huffingtonpost.com/entry/earth-day-eco-fiction-climate-change_us_5ad92237e4b0e4d0715ec872.

Williams, Rowan. "A Ray of Darkness." In *A Ray of Darkness: Sermons and Reflections*, by Rowan Williams. Cambridge, MA: Cowley, 1995.

————. *Resurrection: Interpreting the Easter Gospel.* London: Darton, Longman and Todd, 1982; Cleveland: Pilgrim, 2004.

Williams, William Carlos. *The Collected Poems of William Carlos Williams, Vol. II: 1939–1962,* edited by Christopher MacGowan. New York: New Directions, 1988.

World Economic Forum, Ellen MacArthur Foundation, and McKinsey & Company. *The New Plastics Economy—Rethinking the Future of Plastics.* 2016. http://www.ellenmacarthurfoundation.org/publications

World Health Organization. "Climate Change and Health." February 1, 2018. https://www.who.int/news-room/fact-sheets/detail/climate-change-and-health.

World Meteorological Organization. "WMO Greenhouse Gas Bulletin, No. 15," November 25, 2019. https://reliefweb.int/sites/reliefweb.int/files/resources/GHG-Bulletin-15_en.pdf.

Yale Program on Climate Change Communications. "Climate Change in the American Mind: April 2019." June 27, 2019. https://climatecommunication.yale.edu/publications/climate-change-in-the-american-mind-april-2019/2/.

Index

Page numbers in italic indicate figures.